W9-ACZ-981

WORLD ENERGY STRATEGIES

Facts, Issues, and Options

PUBLISHER'S PREFACE

Friends of the Earth International believes *World Energy Strategies* contains a store of information vital to all who, having wondered which experts and which numbers to trust, wish to play their own responsible role in shaping vital decisions now being made around the earth.

Mr Lovins gives a solid but sprightly treatment to highly technical material. Governments — and their constituencies — might well learn the essence of his message by heart, a message he capsules neatly in alluding to people who await the Second Coming of Prometheus and would do well to plan on something more modest.

Although Mr Lovins usually measures energy in joules, he also speaks of "energy slaves" and cites 12 as the world-wide per capita equivalent of man's present use of energy. The global population, free and slave, thus totals 50 billion; the US, using one third the world's energy, has a total population of 17 billion and is the most populous country on earth.

Such inequity in energy use warrants major consideration in determining world energy strategies. So does the environmental impact of energy. Much of it produces non-biodegradable matter that the environment cannot handle. Much of it reduces protoplasm to heat, which is then wasted. A great deal of it, we are belatedly learning, produces carcinogens, mutagens, and teratogens that distort the essential genetic information that affects all life — in exchange for temporary convenience. Certainly, most of the developed countries' energy slaves are busily borrowing heavily against the future's standard of living to fuel the present standard of using.

Ivan Illich has observed that the proponents of an energy crisis continue to propagate a peculiar vision: that ". . . man is born into prolonged dependence on slaves which he must painfully learn to master. If he does not employ prisoners, then he needs motors to do most of his work." Illich concludes: "The energy crisis focuses concern on the scarcity of fodder for these slaves. I prefer to ask whether free men need them."

Mr Lovins's work, in its earlier versions, has appeared in the UK under the aegis of Earth Resources Research, Ltd., in the US in the *Bulletin of Atomic Scientists*, and under the imprint of Jordens Vänner (a Friends of the Earth affiliate), in Sweden, where it has gone into several printings and has been highly recommended.

DAVID R. BROWER, President
Friends of the Earth International

WORLD ENERGY STRATEGIES

Facts, Issues, and Options

by

Amory B. Lovins

Foreword by Professor Hannes Alfvén
Nobel Laureate in Physics, 1971

Friends of the Earth International
San Francisco • New York • London • Paris
Ballinger Publishing Company, Cambridge, Massachusetts

Friends of the Earth in the United States, and sister organizations of the same name in other countries, are working for the preservation, restoration, and more rational use of the earth. We urge people to make more intensive use of the branches of government that society has set up for itself. Within the limits of support given us, we try to represent the public's interest in the environment before administrative and legislative bodies and in court. We add to, and need, the diversity of the conservation front in its vital effort to build greater respect for the earth and its living resources, including man.

We lobby for this idea. We work closely with our sister organizations abroad, and with new and old conservation organizations here and abroad that have saved so much for all of us to work for.

We publish—books, like this, and larger and smaller—because of our belief in the power of the book and also to help support ourselves. Our environmental newspaper is "Not Man Apart."

If the public press is the fourth estate, perhaps we are the fifth. We speak out for you; we invite your support.

Published in New York by Friends of the Earth, Inc., and simultaneously in London by Friends of the Earth Limited.
Library of Congress catalog number: 75-2010.
International Standard Book Number: 0-913890-09-X.

ACKNOWLEDGMENTS

For their generous contributions, both deliberate and inadvertent, of facts and insights, the author is grateful to Dr Robert Ayres, Mr David Brower, Mr Monte Canfield Jnr, Dr Thomas Cochran, Dr Herman Daly, Dr Warren Donnelly, Professor Hermann Flohn, Professor Dennis Gabor, Mr Henrik Harboe, Dr Carl-Göran Hedén, Mr Paul Hofseth, Dr John Holdren, Lady Jackson, Dr William Kellogg, Professor Henry Kendall, Mr Gerald Leach, Professor Bo Lehnert, Professor Wassily Leontief, Dr Gordon MacNabb, Dr John O'Leary, Mr Walter Patterson, Mr Harry Perry, Dr Jan Prawitz, Dr Walter Orr Roberts, Dr Ignacy Sachs, Dr Frank von Hippel, Dr George Weil, Dr Robert Williams, and above all Professor Carroll Wilson. Other contributors are too numerous to name but none the less valued. Responsibility for all facts and opinions stated in this paper remains that of the author alone.

BIBLIOGRAPHICA

A precursor of this paper was presented by the author at the scientific Symposium on Population, Resources, and Environment, held by the United Nations 26 September—5 October 1973 in Stockholm in preparation for the 1974 World Population Conference. Copyright in that paper ("Energy Resources", document E/CONF 60/SYM III/12) is reserved to the United Nations, which, though it commissioned the paper, notes that the views expressed therein are those of the author and do not necessarily reflect its own.

After extensive revision into approximately its present form, the paper was published as a copyrighted monograph on

15 November 1973 by Earth Resources Research Ltd (9 Poland St, London WIV 3DG, England). The ERR Ltd edition was reprinted in November 1973 and again in early 1974, and has been substituted for the original UN draft in the collected papers of the Stockholm Symposium (to be published by the UN in 1974).

A further revision incorporating substantial new matter was published in the May and June 1974 issues of *Bulletin of the Atomic Scientists* (1020-4 East 58th St, Chicago, Illinois 60637, USA), with the headings and footnotes rearranged.

This edition further revises and updates the *Bulletin* edition and returns to the earlier arrangement of footnotes. In a publication designed for use on both sides of the Atlantic, no orthography can please everyone, but it is hoped that the present compromise (money generally in dollars, spelling and punctuation British except for units of measure) will be reasonably intelligible. Comments and suggestions on style or substance would be most welcome (sent to the London or San Francisco addresses), as the author hopes to continue to improve the manuscript. Several short tutorials—not part of the original manuscript—have been inserted in boxes within the text to help less technically minded readers to understand matters with which they may be imperfectly acquainted.

FOREWORD

The following is a foreword by Professor Hannes Alfven, Nobel laureate in physics, to the Swedish edition of *World Energy Strategies*, by Amory Lovins. That edition was published by Jordens Vänner, the Swedish Friends of the Earth. Professor Alfvén's specific allusions are broadly applicable. This translation of Professor Alfvén's foreword is by Lee Schipper, Information Specialist, Energy and Resources Program, University of California, Berkeley.

Ready access to cheap energy has for a long time been an obvious condition for technological progress, so obvious that energy policy could confidently be left in the hands of experts and kept out of politics. It was only about five years ago that some of the programs advocated by experts were first questioned by scientists, thereby starting an avalanche of discussion that became timely with the oil crisis of autumn 1973. The discussion takes on a special character because the capital invested in the world's energy system amounts to tens or hundreds of billions of dollars, rubles, pounds, yens, marks, francs, and kronor. Besides these investments, there are at least as meaningful commitments of political prestige and military power politics. As a consequence, the energy debate has become a mixture of high-class technical-scientific-social arguments on the one hand and a sales pitch of political and economic character on the other. It is not the actual difficulty of the subject alone that has made and still makes the matter so confusing.

If you peel away the commercial and shortsighted political arguments, however, the international debate begins to clear up. The American physicist Lovins presents here an admirable overview of this debate, which he analyzes in a technically and scientifically competent and intelligent way. Moral and humane viewpoints are given the dominant role that they must play when the future of all of us is at stake. He points out that while we might not be completely sure of what we *ought* to do, the debate has made clear what we *must not* do, if we are to avoid finding ourselves in a dangerous or impossible situation ten years hence. This reasoning receives support from the fact that we were maneuvered into the present crisis by a number of shortsighted measures, and research or political misconceptions, about ten years ago.

For all those who want to study the energy problem, it is important to observe that this really consists of two groups of problems that can advantageously be split up as follows:

1. *The choices among various energy sources.* This is in principle a technical problem — which, however, does not mean that we can leave it to technologists, who often lack the freedom to express themselves owing to political or economic censorship. Perhaps even more important, the various technical alternatives have such profound consequences both for us and for coming generations that it is the right and responsibility of all of us to become involved in what these consequences entail.
2. The other group of problems relate to *how much energy we think we need.* All agree that the present rate of growth of energy use cannot continue forever, and that we must slow down somehow. But which level of use will we settle for? Shall we have a high-energy society or a low-energy society, and what would our standard of living be under either alternative?

From Lovins's analysis, one rightly gets the impression that we can picture reasonable solutions to the first group of problems, but that the second demands a much more profound analysis of what kind of society we really want to have.

Until recently, discussion of energy in Sweden was scanty. The criticism offered from several directions was not answered, but rather ignored. It is only recently, within the past year, that international discussion broke the logjam and it became widely held that Sweden was driving with the accelerator floored into a blind alley. Government and politicians are so strongly bound by established political traditions that a reorientation of these is not in the cards; yet, a highly informed and sufficiently broad public opinion has the strength to make its voice heard.

The energy debate will therefore become a crucial test of democracy, and it is exceptionally important that everyone take a stand. This must not happen simply through quoting of authorities, not even Lovins. The only way to handle a question of this importance is to weigh the various arguments against each other and decide for yourself which are correct and which are false. Lovins provides an excellent guide in the art of this noble debate, and he does not demand to be believed blindly. He instead refers to the presently ample literature for those who will check his arguments in detail and engross themselves in the subject.

—Hannes Alfvén
June 1974

CONTENTS

1

Introduction

INTRODUCTION

Sources
Time-scale
Emphasis
Units

— 1 —

As medical science, by deferring death, has allowed many more people to live on the earth, so the energy of fossil fuels, by deferring physical scarcity, has kept those people alive. Medical technology has caused a population explosion; energy technology, at least for some, a wealth explosion. How many people can continue to live on the earth for how long and with what wealth depends on the ingenuity and wisdom with which man uses energy. Yet on a planet that is round and therefore finite, energy conversion must eventually encounter some geophysical outer limit; and even sooner, it may be constrained by lack of resources, by biological side-effects, by technical problems, or by social, political, and economic pressures.

Exploring these constraints and the ways in which they interact with the numbers and goals of people in widely differing societies is arguably the most complex problem in the world today, and one of the most important. Energy problems are inseparable from problems in almost every area of economic, foreign, industrial, agricultural, and social policy. Any brief survey of world energy strategies must therefore be exploratory rather than definitive, synthetic rather than exhaustive. Furthermore, so many energy problems are controversial, unsolved, or perhaps insoluble that there are few easy or canonically received answers. This summary of a tangled and fast-moving field cannot present a consensus where none exists, but can only suggest where the merits of a dispute may lie, and undertake to represent a respectable body of technical opinion. The aim of this paper is to outline certain assessments and conclusions on which much top-level energy thinking has in recent months begun to converge, presenting such information, and no more, as is needed to convey a sound grasp of the subject. Such an authoritative statement deserves a caveat: never believe an expert. No expert can tell the whole story, nor avoid emotional entanglement with particular ideas. And never, never believe an expert who is trying to sell you something!

Sources

This paper will not obscure the basic outlines of energy problems with a flood of charts, tables, and graphs—partly because words tell most when spiced with data, not smothered, and partly because so many energy data prove on close acquaintance to be inconsistent, unreliable, or irrelevant. A high official of the US Atomic Energy Commission, discussing nuclear power costs, once summarized memorably (if inadvertently) the status of most energy statistics. He remarked:

> "Figures in the literature. . .vary by at least a factor of ten. I am not going to try. . .to give you more accurate. . .figures for three very good reasons: (1) They do not exist even within the Atomic Energy Commission. (2) If they did exist they could not be released for security reasons. (3) If they did exist and if they could be released I wouldn't believe them anyway."

Likewise, since a thorough bibliography would be far longer than this paper, highly selective references will be cited only for specialized statements that might seem especially obscure or controversial, not for those based on standard sources.

The present study draws relatively little on the dozens of energy studies now underway in many countries, for these studies are generally of short-term national fuel policy, not of long-term world energy policy, and tend to neglect such fundamental issues as the role of energy in society. We shall not go so far as to suggest (as some do) that fuel shortages should be relieved by burning energy studies; but the implications of man's energy conversions must be examined from a far wider perspective than that of the technologist, economist, or other specialist alone. So little has this been done that no nation on earth appears to have either a coherent long-range energy strategy or the institutions needed to devise one.

Time-scale

Since this paper is concerned with strategies far more than with tactics, it will consider what must be done now and in future to produce a sustainable energy economy from the mid-1980s onwards, not what ad-hocracy might extricate us from our present acute difficulties. These two-time scales, "strategic" and "emergency", must be carefully distinguished. The major problems that face us today are *ipso facto* insoluble, for it takes time to solve problems, and all we have time for now is Band-Aids; but the problems of the mid-1980s and beyond might be soluble if attacked now. This study will therefore focus directly on how to solve the problems that we shall face after immediate uncertainties have been resolved, and only indirectly on how to muddle through meanwhile. This is perhaps a brutal approach, and in no way intends to underestimate the risks we face over the next decade or so (*e.g.* the risk of further military adventures in the Middle East); but it is a necessary approach if we are to break the cycle of poor planning that has got us into this position. Energy strategy must, in a sense, work backwards: what must we *not* do now if we are to retain the options we shall need later? This approach leads one to conclude that most energy decisions today are based on unsound assumptions and are taking us in an unsustainable direction. This paper will therefore contrast those assumptions with others that could prove more useful.

Power and Energy

Though often used interchangeably by laymen, power and energy are two different ideas to scientists. Energy can do things; power measures how fast energy is doing things. Units of energy measure ability to make things hotter, or higher, or denser, or to change them in other ways; units of power measure the rate at which such things are being done.

Thus a hundred-watt light-bulb is using electrical energy at the rate of a hundred joules of energy per second. If it stays lit for a second, it has used a hundred watt-seconds of energy—that is, a hundred joules of energy. If it stays lit for an hour, it has used a hundred watt-hours, or 360,000 joules. "Using" this electrical energy, however, doesn't make it go away: it is simply changed into light and heat (mostly heat). Energy is never really consumed; it just changes form. It is a law of physics that energy that is doing something tends to be degraded into "lower-grade" energy whose ability to do things is less than before. Energy deteriorates in use; it cannot be completely recycled.

Energy and power can take many forms of which electricity is just one: they can also be mechanical, chemical, thermal (heat), optical, etc. Energy cannot be converted from one form to another with 100% efficiency: some is always converted to less useful, "lower-grade" energy. To make one joule of electricity in a power station, about three joules of heat must be released from chemical or nuclear fuel; two of these three joules are degraded to "waste heat"—the price of making the joule of high-grade electrical energy. A fourth, joule, too, is required to make the one electrical joule's share of the power station! To keep from confusing electrical energy or power with the "primary" source—four times as large—from which it is derived, we must distinguish watts of heat power—watts thermal, or W(t)—from watts of electrical power——watts electric or W(e).

A joule is energy. Energy is often expressed by a heat value—not because all energy is heat, but because so many energy-converting processes involve heat that it's a useful common currency. It takes one thousand fifty-five joules (also called one British Thermal Unit or BTU), applied to a pound of water, to make the water 1°F hotter. This says nothing about how fast the heat is added—that is, how much power is applied to the water. It would take twice as long to heat the water on a 500-watt stove as on a 1000-watt stove, but the total energy added to the water would be the same.

A joule and a watt aren't very big units. Humans are chemical engines running efficiently at 100-150 watts—equivalent to 100-150 joules per second, or about 2000-3000 food calories (kilogram-calories) per day. A horse likes 500 watts—half a thousand watts, or half a kilowatt. Large amounts of power are measured in kilowatts (thousands of watts), megawatts (millions of watts), or gigawatts (thousands of millions of watts). Thus world energy-use is expressed as a large number of small units; joules, British Thermal Units, kilogram-calories, and the like are all small just as a penny is a small, but accurate, unit to use in expressing a year's budget.

In most parts of the world, sunlight falls on a square meter (just over a square yard) of ground at an average rate of hundreds of watts. A square meter of rain forest changes several hundred thousand joules (several hundred kilojoules) of sunlight into plants each day—a power (averaged over day and night) of several watts. A few living communities are slightly more efficient than that; most are much less so. All the plants in the world store sunlight at a total rate of about 50-100 million million watts. The sun radiates energy at a rate of 400,000,000,000,000,000,000,000,000 watts. Half a billionth of that gets to the earth. Of that fraction, about a third bounces off immediately and almost another third remains at the surface to drive the climate. Living things, intertwined by their individually tiny but collectively vast flows of energy, have been evolving for several billion years within the constraints of the energy income available to them. They work very well and know exactly what to do, even though they have never been to engineering school.

Emphasis

Energy studies are traditionally built round a survey of physical inventories of fuel resources, and often lay less stress on constraints of production. This paper will use the opposite emphasis. A resource inventory would by itself be useless in understanding world energy problems, for *energy constraints are not mainly dictated by physical scarcity, but are instead geopolitical, environmental, and sociotechnical* (rate and magnitude problems). Present technology makes it physically possible to convert enough energy to make the earth rather moonlike; this is impressive but irrelevant, since what matters is *how fast and with what political and environmental side-effects* we can meet human energy needs. The nature of these constraints forms the focus of this paper. It also demands four departures from commonly heard economic arguments.

First, discussing fuel availability in purely economic rather than in partly physical terms—discussing reserves to the exclusion of resources, or prices to the exclusion of actual production—is not correct. Fuel minerals, unlike other minerals, are not merely dispersed but physically consumed by combustion: the cost of natural gas is notional once the last molecule of gas has been burned, and before that happens, the price may reflect more the exigencies of current exploitation than the long-term scarcity value of the residual stock. Likewise, the landed price of Libyan crude is irrelevant if the Libyans refuse to ship any: major consumers have just been discovering belatedly that it is sudden real scarcity, not gradually rising price, that requires them to use less oil. Both these divergences of theory from reality show that money and goods are not always equivalent. Though the price mechanism rightly places a high premium on all the tungsten in China, it also places an exceedingly high premium on live dodos.

Second, the price of energy is in a sense less important to the energy planner than to the consumer. Energy prices have

Big numbers

Scientists have a convenient way of writing big numbers without oodles of zeros. It's simple: 1000 becomes 10^3, or $1 \times 10 \times 10 \times 10$, or—as some people prefer to remember it—1 with three zeros after it. Likewise, 2000 is 2×10^3; 3×10^4 is 30,000; 4.6×10^6 is 4,600,000; 10^{-4} is 0.0001 (the decimal point moves four places to the left, not to the right, because the "exponent"—the small raised number attached to the 10—has a minus sign in front of it). To multiply two numbers, add their exponents ($10^8 \times 10^5 = 10^{8+5} = 10^{13}$); to divide two numbers, subtract their exponents ($10^8 \div 10^5 = 10^{8-5} = 10^3$).

If you multiply or divide a number by ten, the result is "an order of magnitude"—a factor of ten—different from what you had: 800 is one order of magnitude larger than 80, and 20 is two orders of magnitude smaller than 2000. Physicists like Mr Lovins like to work in orders of magnitude—often a precise enough style for long-term energy planning because of the great uncertainties inherent in any long-term calculation. Thus if a physicist says "of the order of 3×10^6" he means "somewhere within a range of uncertainty which is an order of magnitude wide and whose center is about 3×10^6"—that is, within the approximate range $1\text{-}10 \times 10^6$. (The range we want is "half an order of magnitude" on either side of 3×10^6, and half an order of magnitude is about a factor of 3.)

Writing in exponential form those numbers which are known only very approximately has a further advantage. Normally, when we write out a big number in full, it is assumed that all the digits shown are "significant", that is, that their exact value is known to be as stated. For example, if we write "126,000", it is assumed that we know we do not mean, let us say, 125,000 or 129,647 or 118,000. The trouble is that we have to write *some* digit in each place whether we know its exact value or not. On the other hand, if we use exponential notation, we can just as well write 1.2 (or 1.3) $\times 10^5$; or, if we are more uncertain than that, 1×10^5 (to show that at least we know the first digit is a 1); or, if we want to be extra sure that people don't read into the number an unwarranted degree of precision, "of the order of 10^5". Thus we can write a number whose order of magnitude we know, without having to suggest that we know its exact value.

little to do with energy costs—especially with external (or, as Garrett Hardin calls them, "larcenous") costs—but are instead fixed mainly by Governments and corporations after their energy policies have been decided on grounds other than future price. Thus including largely fictitious energy prices in economic models of energy flows can lead to circularity.

Third, economists tend to divide energy flows into such "black-box" categories as "supply" and "demand" or "energy sources" and "energy uses". This obscures the physical process—long, continuous chains of energy conversions proceeding from various forms of high-grade energy to low-temperature heat, as required by the Second Law of Thermodynamics. Substituting economic abstractions for thermodynamic realities leads to *e.g.* burning natural gas to raise steam to generate electricity to heat a house, boil water, or dry clothes—all of which could be done directly by the gas 3-6× as efficiently, and none of which is a rational use of such high-grade energy as gas.

Finally, almost every economic analysis forgets that man's energy flows are not mere ciphers on a page, but part of a larger and far more complex network of natural energy flows that are often of similar magnitude. Every change man makes in natural energy flows is a biological and social act. This paper will accordingly be about how the world works, not how we might wish it worked: about the functioning biological world, not an imaginary inanimate one. This approach may help us to learn from the observed working principles of a world that for three billion [10^9] years has been patiently designing stable energy-consuming systems in accordance with physical law.

Units

Units herein are mainly those of the Système International, based on the meter (m), kilogram (kg), and second (s). In this system, the unit of power is the watt (W); your body is now dissipating of the order of a hundred watts of heat. The unit of energy is the joule (J), equal to a power of one watt applied for one second. (Conversely, a watt is a joule per second: 1 W = 1 J/s.) One thousand joules (= 1 kJ) equals 10^{-3}MJ, 10^{10} erg, 0.239 kg-calorie, 0.948 British Thermal Unit, 2.78×10^{-4} kilowatt-hour (thermal) (kW-h(t)), or approximately 3.4×10^{-5} kg of coal equivalent. Conversely, 1 metric ton (1000 kg) of coal, or about 0.7 metric ton of oil, is equivalent to about 8.1 megawatt-hour (thermal) (MW-h(t)) or about 2.9×10^{10} J. These interconvertible units unfortunately say nothing about energy *quality*.

Energy Units

The energy literature is a jungle of both familiar and exotic energy units—joules, BTUs, kilogram-calories (also called kilocalories or kcal), metric tons of coal equivalent, kilowatt-hours thermal, Qs (1Q = 10^{15} or 10^{18} BTU), barrels of oil equivalent, acre-feet[2] (a unit used by some hydroelectric engineers), grams of fissile isotopes, ergs, and even energy slaves. The joule and its associated power unit, the watt, are both internationally accepted metric units with easily convertible values: for example, a BTU is within a few percent of being a kilojoule (1000 joules); and this in turn is about a quarter of a kilogram-calorie or food calorie (the kind used to measure diets). The energy content of fuels such as oil, coal, and uranium is the theoretical content: somewhat less is obtained through actual combustion or fission owing to various practical inefficiencies. The average US rate of energy consumption is about 80 pounds of coal equivalent per day. One barrel of oil equivalent (42 US gallons, weighing about a seventh of a metric ton) contains 5.8×10^9 joules or about 5.5×10^6 BTU. The energy in one US gallon of oil is equivalent to one and a half weeks of a fine diet of 3000 food calories per day (more than most of the world's people get). The gallon lasts less than ten minutes in a fast car. A Concorde SST consumes it in about a tenth of a second. Millenia were consumed in putting the gallon together.

2

World Energy Conversion

WORLD ENERGY CONVERSION

Supply patterns
Demand prospects
Possible rates of innovation

— 2 —

Estimates of the present rate of man's global energy conversion disagree by $\pm 25\%$, but most people would agree on 8×10^{12}W as a round number. This is more than 20× the energy represented by a FAO-standard diet for the world's population. Thus global industry already uses about 20× as much energy as is recovered from all agriculture and hunting on both land and sea, and gives everyone the equivalent of about 12 hardworking slaves (each 175 W ~ 3600 kg-cal/day)—or 50 slaves if we consider their work output in a 40-hour week (at 100% efficiency), rather than their food input in a 168-hour week. In density of power per unit of continental area, human energy conversion is now about 1/4 as great as net photosynthesis and is roughly equal to the natural outward flux of geothermal heat from the earth's interior[1].

On world average, man's energy conversion has recently been increasing almost 3× as fast as population, *i.e.* at about 5.7%/yr,* equivalent to a 4.5× increase by the year 2000. Annual growth is typically[2] 1-3 percentage points more rapid in poor countries than in, say, the USA (just over 4%/yr)—which, however, with less than 6% of world population, uses over 1/3 of world energy, or roughly *twice* the combined total for Africa, the rest of North and South America, and Asia except Japan.

[1] Study of Man's Impact on Climate (SMIC), *Inadvertent Climate Modification,* MIT Press (Cambridge, Massachusetts), 1971.

[2] There are many exceptions, principally in countries with severe population pressures; *e.g.* per-capita energy conversion is growing about 1/4 as fast in India as in the USA, 1/5 as fast as in the EEC.

* This growth is in *primary* energy, not in *end-use* energy. The difference is important. Leach calculates that since 1900, gross energy conversion in the UK has doubled while *energy at the point of final use has only increased by half* (or by a third per capita). The difference has been swallowed up by the fuel industries, owing to electrification and the switch from coal to oil (which requires more energy to produce).

This extreme inequality of energy distribution is among the most striking and disquieting features of today's energy picture. Gaps of one or even two orders of magnitude are common: *e.g.* there is a 250× per-capita energy gap between the USA and Nigeria. (Likewise, some 94% of human energy conversion is in the Northern Hemisphere, and 75% is in perhaps 0.3% of the earth's land area[1].) Yet whatever correlation may exist between per-capita energy conversion and various social indicators is controversial, tenuous, and non-linear. Few Americans, for example, would argue that they live twice as well as the English or three times as well as the French, in proportion to their per-capita energy conversion. Even correlations of per-capita GNP (hardly a reliable social indicator) with per-capita energy conversion show very large scatter from country to country.

In some of the nations with the grossest national products, energy conversion per unit of GNP is increasing, owing to changes in technologies (*e.g.* from natural to synthetic materials) and to such market shifts as electrification—heavily encouraged through advertising, promotional rates, and other tactics by an energy industry that tends to identify the national interest with its own. (In the UK, one arm of the nationalized electricity industry will cheerfully sell a £7 electric fire which commits another arm to installing more than £200 worth of capacity—but which successfully competes with similar promotional efforts by the nationalized gas industry.) Energy promotion has led to widespread confusion between demand and need. It also supports self-fulfilling prophecy, whereby energy planners first predict a doubling of, say, electrical production in ten years and then work hard to make it come true—and complain about how difficult it is, perhaps simultaneously proclaiming "Energy crisis!" and "Use more!" The world-wide rush to acute epidemic energitis reminds many thoughtful observers of Santayana's remark: "Fanaticism consists in redoubling your efforts when you have forgotten your aim." As assumed relationships of cause and necessity between energy conversion and social welfare have become widely accepted, energy conversion seems to have become a social good to be maximized.

The profligacy of industrial societies is founded on a pervasive attitude: that energy is a virtually free good that can be substituted for all other forms of capital. (We shall study some examples below.) But in most of the world, necessity enforces a very different attitude: that energy is an expensive good which one can hardly afford to substitute for anything, and which one will be even less able to afford as others' extravagance exhausts cheap reserves and drives up the price. Countries with expanding industry, rapid population growth (often with skewed age-structure), and a trend towards intensive agriculture will be hit especially hard by economic energy scarcities from now on. It is these countries that rising energy prices will most damage—not the industrial countries for which far more expensive energy is long overdue.

Supply Patterns

Though no commercial source of high-grade energy can be considered in isolation, the basic global statistics of major sources[3] are easily summarized in approximate and aggregated form:

a) About 97% of primary energy comes from fossil fuels, which are nonrenewable except in geological time. This overwhelming commitment comprises:
   1) about 38% solid fuels (consumption rising relatively slowly and with fluctuations);
   2) about 40% oil (consumption doubling about every decade through 1973);
   3) about 19% natural gas (consumption doubling about every 7-8 yr).

[3] We are speaking here, of course, of industrial energy, and must not forget that in many countries, the main source of energy is people. People are extremely efficient users of food (if they get enough), and the food itself takes little energy to grow (*infra*) if energy-intensive agriculture has not yet taken over from traditional methods. Those who believe that nothing substantial can be done without machines and heavy industry should examine the People's Republic of China.

b) Hydroelectricity provides only about 2% of world primary energy, and geothermal energy far less.
c) Nuclear fission, wood, and other sources probably account for less than 1% each.
d) This global pattern masks wide regional and national variations: *e.g.* the USA uses nearly 2/3 of world gas; Japan uses almost twice as much oil as coal, but almost no gas; the USSR and Communist Asia rely heavily on coal; many of the less industrialized countries have disproportionate hydroelectric production, and some burn substantial amounts of vegetation or animal wastes[4] for fuel.

Demand Prospects

Before we critically assess the outlook for these and other sources of primary energy, it is important to observe that many proposed technologies in a "full-steam-ahead" economy would make the rate of increase of world energy conversion increase even faster than it appears to have done recently. For example, modern (as opposed to traditional) agriculture already requires fossil-fuel subsidies so heavy that we must often supply several times as much chemical and mechanical energy as we recover in food—according to one estimate[5], about 5× as much in all US food production, compared with 1/20 – 1/50× in typical "primitive" agriculture[6]. But this is only the beginning:
a) For world food yields to keep pace with the population growth which will occur if mortality rates do not greatly increase (owing to the failure of food supply to keep pace),

[4] India is said to burn 1-3 $\times$ 10^{11} kg/yr of dung (*cf.* Britain's coal consumption of 1 $\times$ 10^{11} kg/yr). The residual ash, along with a comparable amount of unburned dung, is used as fertilizer.

[5] Perelman, J J, *Environment 14,* 8, 8 (1972). Leach's more sophisticated analysis[7] yields similar results; *cf.* Steinhart, J S and C E, *Science 184:*307 (1974).

[6] Black, J N, *Proc Assoc Appl Biologists* 67:272 (1971).

production of nitrate fertilizer must increase nearly 100× in this century. (This is consistent with observed diminishing returns in nitrogen takeup.) Present technology—recently much improved, and probably hard to improve much further—demands about 2.3 kg of coal equivalent per kg of fixed nitrogen[7]. (A breakthrough in microbiological nitrogen fixation may or may not reduce this cost.) Such energy costs imply that nitrogen-fertilizer factories alone in the year 2000, to produce $\sim 8 \times 10^{11}$ kg nitrogen/yr, may require about 20% of present total world energy use.

b) Desalinating seawater requires 3 MJ/m^3 in theory[8], about 170 in present practice, and (speculatively) about 50 in very-large-scale future practice[9,10]. The energy requirements of extensive desalination are thus impressive. For example, under reasonable assumptions[11], desalinating enough seawater to grow one average man's subsistence crops could well require as much energy as he now uses for everything. This is to say nothing of capital, rate and magnitude, or environmental problems.

c) Metals further illustrate the energy-intensiveness of cornucopian technologies: producing 1 kg of aluminium from

[7] Leach, G, "The Energy Costs of Food Production", in Bourne, A, ed, *The Man-Food Equation,* Academic Press (London), 1974; preliminary energy network analyses, among the best available. *Cf.* Pimentel, D, *et al, Science 182:*443 (1973), and Slesser, M, *J Sci Fd Agric 24:*1193 (1973).

[8] This calculation of the thermodynamic minimum energy is based on the osmotic pressure of salt, derived from the ideal gas law; it assumes *reversible* desalination—hence much of the gap between theory and practice. Calculation of the entropy of mixing of salt in seawater yields a similar result.

[9] Weinberg, A M and Hammond, R P, *Am Scient 58:*412 (1970).

[10] Weinberg, A M, *Bull Atom Scient 26,* 6, 69 (1970).

[11] We assume here that a mean flow of 1 m^3/day, desalinated by 100 MJ/m^3 and pumped by 20 MJ/m^3, will grow (after evaporative losses) 2500 net recoverable kg-cal/day of protein-rich crops. These crops are assumed to form 90% of the diet, and the remainder is assumed to be derived from the crops at 5:1 conversion.

bauxite requires 18 MJ in thermodynamic theory and more than 10× as much in practice—a few kW-h of total primary energy per beer-can. (The doubling time of world aluminium production—now about 10^{10} kg/yr—is less than a decade.) It is less commonly realized that metals with lower binding energies but higher entropy can be nearly as energy-intensive: just mining and beneficiating ore yielding 1 kg of copper, for example, takes about 50 MJ[12, 13]. Total direct energy inputs[14] for producing certain metals are:

copper from 1.0% sulphide ore in place (1940s)	54 MJ/kg
copper from 0.3% sulphide ore in place (1980s)	98 "
aluminium from 50% bauxite in place (1970s)	204 "
magnesium from seawater (anytime)	360 "
titanium from ilmenite in place (1970s)	593 "
(*cf.* combustion of coal, which yields)	29 "

Energy costs for leaner sources will be higher: in particular, comminution (crushing and grinding) of typical nonferrous ores to liberation size now requires about 50-80 kJ/kg ore[13], implying total production energies of the order of 10^{10} J/kg metal in a 10^{-5} "ore". This enormous energy cost is important because it appears likely that for most nonstructural metals there is not a continuum of progressively poorer ores to be mined at steadily increasing cost in energy and land, but rather an abrupt

[12] OECD, *Gaps in Technology: Non-ferrous Metals,* OECD, Paris, 1969.

[13] Lovins, A B, *Openpit Mining* (Earth Island, London), 1973. Available from Friends of the Earth Ltd, 9 Poland St., London WIV 3DG.

[14] Bravard, J C, Flora, H B II, and Portal, C, "Energy Expenditures Associated with the Production and Recycle of Metals", ORNL-NSF-EP-24, Oak Ridge National Laboratory, November 1972. Transport costs for both products and raw materials are neglected; the assumed conversion efficiency of primary fuel to electricity is 40%. A more complete analysis of direct and indirect inputs, done by P F Chapman of the Open University (Milton Keynes, Bucks., England), yields results typically 40-100% higher.

grade gap, often[15] of 10^3–$10^4\times$, between mineralized and barren rock. Thus for some such metals, mining energy may rise by orders of magnitude several decades hence.

One of the few bright spots in this outlook is that, contrary to what energy promoters often suggest, technologies for abating pollution and for recycling scarce materials are rarely energy-intensive, and often yield net energy savings[16, 17]

In summary, the present trend towards substituting energy for other resources—rich ores, fresh water, highly productive ecosystems, etc—suggests that if energy supply were not constrained in any way, demand would rise very steeply. In such circumstances one can easily envisage, by late in this century, a doubling of current growth rates. However, this is not going to happen, and indeed the *present* energy growth rate cannot be maintained for long[18]. To explain why this is so, we must now examine the outlook both for fossil-fuel supply and for those new energy sources which prophets of the Second Coming of Prometheus believe will prevent scarcity. We shall then discuss alternative strategies and the issues they raise, including climatic constraints.

Possible Rates of Innovation

To place in perspective the supply problems of fossil fuels, we must stress a basic insight more common amongst engineers

[15] Lovering, T S, "Mineral Resources from the Land", in National Academy of Sciences / National Research Council, *Resources and Man,* W H Freeman, San Francisco, 1969.

[16] Hirst, E, *Environment 15,* 8, 37 (1973).

[17] Perry, H, "Conservation of Energy", Committee on Interior and Insular Affairs, US Senate, Serial 92-18 (USGPO 5270-01602), 1972.

[18] The argument, implicit in much of this paper, that all energy forecasts are probably wrong may seem a well-aimed boomerang if followed by an energy forecast. It is easier, however, to forecast correctly what will *not* happen than what *will* happen.

than amongst politicians: the truly formidable rate and magnitude problems of *any* major innovation. The aggregate amounts of energy now being converted are so prodigious that voluntary rapid change in supply patterns is *physically impossible.* For example, suppose that our present world conversion rate of 8×10^{12} W—97% of it from fossil fuels—continues to grow (as most authorities predict and urge) by about 5%/yr for the rest of this century, yielding a 3.7× increase to about 3×10^{13} W. If we could somehow build one huge (1 GW = 1000 MW(e) = 10^9 W(e)) nuclear power station per *day* for the rest of this century, starting today, then when we had finished, *more than half* of our primary energy would still come from fossil fuels, which would be consumed about *twice* as fast as now. Few knowledgeable people would say that such a rapid nuclear infusion is possible, even were it advisable. On the other hand, such rapid sustained growth in energy conversion as we have just assumed will not actually occur, owing both to supply constraints and to the moderation in demand which these constraints will encourage and compel.

3

Oil

OIL

Exploration
Costs
Foreign and fiscal policy
Recovery
Conclusions

—3—

The pattern of constraints emerges most clearly for oil, the main source of fossil energy capital (10^8-year-old sunlight) being consumed today—and a source whose total effective lifetime will probably be 70-80 years[1]. The most important single datum here is that roughly 2/3 of the world's ultimately recoverable liquid oil resources, and more than half of current reserves, occur in the unique Persian Gulf region. (Most of the rest is divided amongst the USSR, the Western Hemisphere, and to a lesser extent Africa.)

Exploration

Geological knowledge no longer justifies the hope that new discoveries, though they will be extensive and sometimes tactically important, will significantly alter this strategic picture. The conditions necessary for the occurrence of oil, and the location of the earth's main sedimentary basins, are now sufficiently well-known for total recoverable resources to be estimated with some confidence[1]. The substantial deposits which remain to be proven will not greatly shift the time of ultimate physical depletion (*viz.* in the first third of the next century, assuming that consumption continues to grow fairly rapidly for the next 15-20 years as projected by virtually all Governments): if oil consumption continues to double every decade, then *a doubling of world reserves will delay depletion by only a decade.*

[1] Hubbert, M K, "Energy Resources", in National Academy of Sciences / National Research Council, *Resources and Man,* W H Freeman (San Francisco), 1969. *Cf.* Gillette, R, *Science 185:* 127 (1974).

Indeed, to keep the ratio of reserves to production from falling, as it has done for the past five years or so, we ought now to be finding every year several new deposits each comparable to the proven reserves of the North Sea or of the Alaska North Slope. This is not happening, and there seems no likelihood whatever that technical progress will make it happen, for undiscovered oil is now scarce. Since the easier areas outside the well-known producing provinces have already been explored, most new exploration is now in remote, hostile, and fragile environments (such as the Arctic and the Outer Continental Shelf). Despite great effort and ingenuity, US discovery rates[2] have fallen steadily from over 800 barrels[3] per meter drilled in the 1930s to perhaps 100 today[1], and similarly abroad. Accordingly, prominent exploration experts have recently predicted that total world production of liquid oil will peak by about the end of this decade—or a few years later if production does not rise much—and will decline thereafter.

Some data about the widely acclaimed North Sea basin may help to put new discoveries into perspective. The entire recoverable North Sea resources, once proven, are expected to amount to some 1.5-1.8% of ultimately recoverable world resources. North Sea oil is costly to discover (about $5 million per 4000-m wildcat hole) and to extract (about $100 million to build a single platform producing 10^5 barrel/day, plus about $40,000/day to run it). It will cost at least $15-25 billion, attracted from all over the world, to produce the North Sea's likely total capacity of perhaps 6-8 million barrel/day (3-4 $\times$ 10^8 ton/yr) in the mid-1980s; yet this is not enough oil to cover

[2] These figures are somewhat distorted by institutional forces, including peculiar US tax laws which can encourage unsuccessful drilling. Nonetheless, the world ratio of reserves to production has been falling for some years and shows no sign of recovery.

[3] Approximately 7 barrels (the customary unit) of crude oil weigh 1 metric ton. One barrel is thus approximately 5.8 GJ = 5.8×10^9 J. OECD statistics commonly use the large unit of 1 million metric tons of oil equivalent, nearly equal to 10^{13} kg-cal.

the traditionally projected *increase*[4] in Western Europe's oil consumption (now about 15 million barrel/day and doubling recently every 8-10 years[5]. The required offshore technologies, too, meet or exceed the safe limits of those now available[6], creating unknown but possibly great environmental risks and the likelihood that the social disruption[7] of sudden industrialization will be increased by an unfavourable tradeoff between the oil and fishing industries[8].

Costs

The accompanying table, concentrating on broad ranges rather than on technical details, will summarize more generally the *approximate* 1972-73 costs of producing crude oil or its energy equivalent from various areas or methods—both the typical capital cost of 1 barrel/day capacity (excluding escalation and interest) and the technical unit cost of a barrel at the wellhead (including exploration and lifting costs but excluding carriage, taxes, producing Government's economic rent, and producing company's profit).

[4] The projections are dubious, however, as they generally imply that oil consumption will soon have surpassed total energy conversion.

[5] It is not obvious, however, that anything short of Draconian controls can keep much of the oil in Europe: trans-shipment to the USA is a likely result of US-Japanese competition for low-sulphur crudes. The foreign sources of most of the North Sea capital are unlikely to welcome export restrictions.

[6] Sanders, N, *New Scientist* 56:380 (10 November 1972). Pipelines or no, it seems likely that much of the oil will come ashore by itself whether we want it or not. *Cf.* Kash, D E *et al, Energy Under the Ocean* (U. of Okla. Press, Norman, Okla., 1973).

[7] Francis, J and Swan, N, *Scotland in Turmoil,* Church of Scotland Home Board (Edinburgh, January 1973).

[8] Many would argue that the social cost of hasty oil development can be even higher: in Alaska, for example, the extinction of a native culture.

Approximate Costs of Producing Crude Oil or Its Energy Equivalent, 1972–1973

Energy Source	Capital Cost ($/[bbl/day])	Technical Unit Cost ($/bbl)
Persian Gulf	100–300	0.10–0.20
Nigeria	600–800	0.40–0.60
Venezuela*, Far East, Australia	700–1,000	0.40–0.60
North Sea, most other Europe	2,500–4,000	0.90–2.00
Large deep-sea reservoirs	over 3,000?	2.00– ?
New US reservoirs (not too remote)	3,000–4,000	1.70–2.50
Easier part of Alberta tar sands	4,000–8,000	2.00-6.00
High-grade oil shales	4,500–9,000	3.00–4.50
Gas synthesized from coal	5,500–8,000	3.00–6.00
Liquid synthesized from coal	6,000–9,500	3.00–6.00
Liquified natural gas (landed)	6,000–10,000	3.00–6.00
Nuclear fission (light-water reactor)	20,000–30,000	?

* Excluding heavy oils.

The table makes clear the enormous leverage exerted by holders of large reserves at low technical unit cost. It is this virtual monopoly, not on energy, but on cheap energy, that is rapidly bringing the era of cheap energy to an end[9]: from now on it is a

[9] This has somehow come as a surprise to many: even in February 1970, the US Cabinet Task Force on Oil Import Control could report that "we

matter of what the market will bear, and oil in the ground will appreciate faster in free storage than would invested oil revenues. Even more important, major oil producers[10] no longer feel obliged, morally or otherwise, nor have they incentives, to produce oil at a rate clearly detrimental to their economies. Such considerations helped, even before the 1973 war in the Middle East, to lead Libya and Kuwait to impose production ceilings; others may follow suit. It now seems most unlikely that sufficient sustained increments of production will actually occur to meet 1973 forecasts of world oil demand over the next few decades—a fact belatedly admitted in private late-1974 forecasts by some international organizations.

Foreign and Fiscal Policy

The world oil market has been made especially volatile by the USA's rapid emergence as a major oil importer in direct competition with other buyers, both rich and poor—most notably Japan and the EEC, both already absolutely dependent on Middle Eastern oil. With US oil and gas resources at or past their production peak and heading for virtual exhaustion by the turn of the century[1], many US planners have long estimated that the US would import about 10 million barrel/day in 1980 and twice that much (1/4 of her total energy) in the mid-1980s. Such projections assumed rapidly growing Middle Eastern

---

do not predict a substantial price rise in world oil markets over the coming decade." (The price was then $2/bbl.) In Britain, documents released by the Central Electricity Generating Board in the spring of 1973 revealed the assumption that fuel costs would rise a few tens of percent to 1980 and then remain constant to 2010. (Lovins, A B, *New Scientist* 58:564 (31 May 1973)).

[10] Khene, A, *Petrol Times,* 16 November 1973, p 4. There is a parallel with current political trends in both Canadian Provinces and US States that export oil to their neighbours.

productive capacities, both technical and political[11]. It was also appreciated, though only vaguely in some quarters, that heavy US importation would give the producing countries immense dollar liquidity[12] which would permit disruptive short-term capital flows, price wars, heavy spending on armaments[13], and extensive downstream investment. Dependence mainly on the few countries (Saudi Arabia, Iran, etc) best able to expand production would also increase political risks.

Recent events, while demonstrating the naïveté of ignoring such risks, have brought into sharp focus the difficulty that major oil consumers face in devising a form of payment which they can maintain and which oil producers will consider an attractive asset. The World Bank has calculated[14] that if oil costs $8/barrel and world oil demand increases modestly, the five nations best able to supply incremental demand and least able to absorb revenues[12] (Saudi Arabia, Abu Dhabi, Qatar, Kuwait, Libya) will in 1980 have accumulated roughly 3/4 of world liquidity. That plainly cannot happen without severe disruption of national economies and of world trade; but more plausible alternatives[15] seem limited. Price rollbacks would probably lead to a similar result less than a decade later.

[11] Fells, I, *et al,* "Energy for the Future", The Institute of Fuel, 18 Devonshire St, London WlN 2AU, 1973; "Foreign Policy Implications of the Energy Crisis", Hearings, Subcommittee on Foreign Economic Policy, Committee on Foreign Affairs, US House of Representatives, 1972.

[12] Akins, J E, *Foreign Affairs,* April 1973.

[13] Among the hidden costs of Middle Eastern oil is the cost of the weapons "needed" to "protect" it—and in turn to counter the destabilizing effect of *those* weapons.

[14] "The Implications of Rising Petroleum Prices" (1973); summarized in *Weekly Energy Report* (Washington, DC), 14 January 1974, p 6, and later revised and expanded by the World Bank.

[15] Cynics might identify three main outcomes: (1) Middle Eastern oil is shipped as consumers desire, resulting in excess producer liquidity[12], local economic and political instability, and war. (2) Oil shipments are restricted, leading to economic disruption of, and Middle Eastern military adventures by, consumers, and hence to war. (3) Consumers strive to ensure that they won't care whether they get the oil or not; for Western Europe and Japan, that day is far off. In *Foreign Affairs* 52:690

Medium- and long-term downstream investment is risky for both parties, is therefore nearing saturation already, and returns embarrassing revenues. Industrial barter is limited by (*inter alia*) oil producers' infrastructure. Short-term investment incurs risks from devaluation, blocked conversion, and inflation. The problems of scale are immense: neither all US assets overseas nor the annual world armaments budget could pay for a year's oil in the early 1980s. Thus rich countries face the novel problem of being unable to pay money for oil to countries which already have most of the stocks of both. Poor countries, less able to buy their way out of trouble and more dependent on oil-based fertilizers, are even more vulnerable; and recycling oil revenues as development aid, even if it would suffice, seems politically unlikely. These balance-of-payments problems will also probably be aggravated by imitative rises in the prices of other raw materials as various exporters discover that they too can earn more by producing less. No doubt new ways of recycling capital to relieve such problems will be devised, and such innovations will work for a few years until they don't work any more; but tne scale of the innovations required makes their long-term success doubtful. Carroll Wilson's ingenious proposal (oil revenues to be recycled to oil consumers in proportion to their success in meeting annually reviewed goals in energy and oil conservation) may stand the best chance despite substantial practical difficulties. *

Perception of the risk both of oil producers' withholding supplies and of consumers' inability to buy them may encourage strategic stockpiling. This is awkward, perhaps hazardous, costly (several dollars per barrel), and dependent on surplus oil

---

(July 1974), W J Levy summarizes why the problem is so difficult. In effect, oil producers do not want Monopoly money; oil consumers therefore have an array of possible methods of going in hock by transferring more attractive assets; some of these methods retain an option of later expropriation while others do not; and both producers and consumers know this to be the position.

* Wilson's joint paper with Farmanfarmaian, K *et al*, *Foreign Affairs*, *53*:201 (January 1975) sets out some intriguing new proposals.

imports. Its maximum useful life is not much greater than Western Europe's 2-3 months; whereas such logistical problems as shortages of fabricated steel (pipelines, etc) and of steel production capacity could easily extend major supply interruptions of military origin to over a year. Political and moral issues aside, therefore, further Middle Eastern hostilities (whether by invasion or by proxy) carry worldwide risks whose worldwide importance may not be widely enough appreciated.

The economic risk run by companies investing in such capital-intensive areas as the North Sea while other areas contain tens of times as much oil at a tenth the technical unit cost may be substantial—especially because only large fields (similar to those that now contain nearly 3/4 of current reserves) can support the high cost of exploration and production in such areas. Nonetheless, countries which believe they can retain access to offshore reserves are almost without exception planning massive investment and immediate rapid extraction; *e.g.* North Sea oil production from proven fields in the Norwegian sector is to peak in 1977, decline to half-maximum within about four more years, and decline further to almost zero by 1990[16]. (Gas production is to peak 1980–7.) The cogent arguments[17] for

[16] These figures for the combined production from Ekofisk, West Ekofisk, Edda, Eldfisk, Cod and Tor (perhaps 1/5 of total Norwegian resources south of 62°N lat) are derived from a graph in the Norwegian Government report no. 51 (1972–3) presented as part of a Ministerial briefing on oil landings. They have since been confirmed independently. Norwegian policy, however is now belatedly swinging toward delayed and stretched production, and even further exploration may be severely restricted on the grounds that anyone who discovers oil is already sucked into the whirlpool of extraction. The Ministry of Finance White Paper of 15 February 1974 (Storting report 25) proposes a strict "go-slow" policy.

[17] Those not yet mentioned include retaining a bargaining counter (North Sea oil no longer fills this role once extraction has begun), saving domestic reserves for the time when world prices are highest and suppliers least amiable, minimizing social and environmental disruption, evolving safer technologies and a healthy indigenous support industry, reducing the duplication of costly facilities and the temptation

delayed and stretched production seem to carry no weight with Governments whose resource decisions are apparently made by accountants on the basis of discounted cash flows.

Recovery

Only about 30–35% of the oil in place in an average reservoir is now recovered, compared with about 80% for gas, 55% for underground-mined coal, and 80% for surface-mined coal[18]. As prices double or treble, intensified secondary and tertiary recovery may be expected to increase oil recovery by 5–10 percentage points; recovery rates have risen by about 10 points in the past two decades. In some cases (*e.g.* Venezuelan heavy oils) improvements may be greater. Recoveries of the order of 50–60% are not expected to become common, however, if only because the amount of energy going down the borehole as steam and the like soon exceeds the energy recovered as oil. Substantial technical breakthroughs in low-cost recovery techniques seem unlikely.

Conclusions

We have deliberately ignored many complexities of oil, including the blending requirements of refineries, the relationships of demand for various refined products, the methods of refining various crudes into various proportions of different products, the

---

to put them in inappropriate places, developing needed technical and managerial expertise at home, and stretching social benefits. The two critical questions to ask are: Is oil in the ground our best investment? What shall we do for an encore? It is unfortunate that in some countries, such as the UK, balance-of-payments pressures (due partly to prolonged failure to cure energy waste) are now so overwhelming that such questions are considered to be of purely academic interest.

[18] Perry, H, "Conservation of Energy", Committee on Interior and Insular Affairs, US Senate, Serial 92-18 (USGPO 5270-01620), 1972.

public-health constraints imposed by the sulphur and trace-element content of various crudes, the uncertain status of desulphurizing and stack-gas-cleaning technologies (now being rapidly deployed in Japan[19]), the growing difficulty of siting refineries in most industrialized countries, the consequent industrialization of *e.g.* the Caribbean, the cost and side-effects of marine transport[20] through congested shipping lanes (oil is the world's largest shipping commodity by tonnage), and the possibility that oil may become more important for its molecular structure than for its heat content. These complexities do not, however, affect the general structure of the world oil problem: cheap oil, for long far too cheap, has now become so highly localized that it will not remain cheap—nor, on a time-scale of a few decades, plentiful. The landed prices of Middle Eastern crudes in the EEC and USA, already rising steeply even before the 1973 war, could rise far beyond $10/barrel within the next decade, and production will probably not meet consumers' wishes. Poor countries struggling to expand nascent industry will be most seriously affected. Such high prices do, of course, open up many newly competitive options, but these take so long to develop and deploy widely that oil prices will go far to determine energy prices no matter what cheaper alternatives are available on paper. Some industrial countries may try simply to pass their extra energy costs on to others, not merely by competitive devaluation but by increasing their industrial output—a sort of reverse price elasticity of demand (viable only if all countries could be simultaneous net exporters, or importers, of the same products).

[19] "Report and Conclusions of the Joint Ad Hoc Group on Air Pollution from Fuel Combustion in Stationary Sources", Environment Directorate, OECD, Paris, 1973. J T Dunham *et al* review recent US progress in *Science 184:*346 (1974). Despite vocal protests by electrical utilities, the US Environmental Protection Agency has found in its thorough *National Public Hearings on Power Plan Compliance with Sulfur Oxide Air Pollution Regulations* (January 1974) that effective and reliable stack-gas-cleaning technologies are now available at reasonable cost.

[20] Mostert, N, "Profiles: Supertankers", *The New Yorker,* 13 and 20 May 1974.

4

Other Fossil Fuels

OTHER FOSSIL FUELS

Natural gas
  Commitments to premium fuels
Coal
  Side-effects
  Fuel policy
Tar sands
Oil shales
  Comparison with coal

— 4 —

Natural Gas

Unlike oil, natural gas is an exceptionally clean and convenient premium fuel whose use for low-grade purposes, *e.g.* raising steam, will probably be banned in many countries within a few years. The energy content of ultimately recoverable world resources is of the same order for gas as for oil, with which it tends to be geologically associated (though decreasingly so as gas exploration techniques improve). Cheap reserves of gas may be slightly closer to depletion than those of oil, however, because growth in demand, despite a somewhat later start, has been more rapid. (In the USA, for example, a very low fixed gas price stimulated a huge artificial market which soon encountered, several years ago, the practical limits of domestic production rates. Gas now provides 1/3 of US energy and is the sixth largest US industry[1].) The USSR apparently holds about 1/3 of ultimately recoverable world gas resources, and North America and the Middle East about 1/5 each[2].

Gas is far more costly to transport overseas than oil. Massive efforts are now underway to augment pipelines and the like with liquefied natural gas (LNG) marine carriers, so as to export more gas from surplus to deficit areas (mainly Japan and the USA) and to help eliminate the flaring that still wastes a substantial fraction of gas production. We shall examine later this extremely expensive and demanding technology, and only

[1] Freeman, S D, *Bull Atom Scient 27,* 8, 8 (1971).

[2] We have not explicitly treated here the natural-gas liquids produced together with natural gas and subject to similar supply constraints. Consisting of *n*-paraffins with more than two carbon atoms, they are generally about 1/5 as plentiful as crude oil, with which they may be aggregated as "total petroleum liquids".

note now that its advocates expect it to grow 40-45%/yr during this decade and over 7%/yr thereafter.

Commitments to premium fuels. Short-term planning by major petroleum consumers has led to heavy and hard-to-reverse commitments of capital: *e.g.* some 200 million cars, each burning roughly its own mass in fuel each year with about 10-20% efficiency[1, 3]. This world car fleet, officially projected[4] to double by 1985, consumes as fuel about 6% of all world energy (2/3 of it in the USA, where the cars are twice as big as elsewhere) and about 1/8 of world oil production (close to half of US crude-oil consumption)[4]. The total energy cost of cars[5], including *inter alia* roads, factories, sales, and maintenance, is nearly twice as large. Such a politically sensitive commitment, reflected in urban and social dependence, conflicts with resource constraints that will become acute within the lifetime of cars now on the road, and with other politically sensitive side-effects such as marine oilspills, refinery nuisance, and air pollution. Yet the dangers of short-term transport planning still seem remote to many Governments—even in the EEC and Japan, which currently depend on imported oil (mostly from the Middle East) for about 2/3 of their energy, but are making little effort in the essential task of restricting premium fuels to nonsubstitutable uses.

Short-term fuel planning can be very costly, not least in the lifetime of associated capital. As Carroll Wilson writes[6]:

> The painful fact is that some part of total refinery capacity, and as much as half of the planned superport capacity, will

[3] Perry, H, "Conservation of Energy", Committee on Interior and Insular Affairs, US Senate, Serial 92–18 (USGPO 5270–01602), 1972,

[4] Leach, G, "The Motor Car and Natural Resources", OECD, Paris, 1972.

[5] Input-output analysis of US energy in 1963 ascribes to cars more than 20% (Herendeen, R A, Center for Advanced Computation document no. 69, University of Illinois, Urbana). Estimates of the 1973 figure range upwards by as much as 2×.

[6] Wilson, C L, "A Plan for Energy Independence", *Foreign Affairs,* July 1973.

become obsolete or surplus as we carry out the shift to coal and the reduction in the proportion of our energy needs supplied by oil imports. When a problem has been neglected as long as we have ignored or misjudged the energy situation, the short-term requirements may not mesh with the requirements for the medium and longer term—and so it is in this case. We have to fix the roof and build a new house at the same time.

Coal

Fortunately, world fossil-fuel resources are not limited to oil and gas. Coal and lignite resources an order of magnitude larger—2/3 of them in Asia—await proving and extraction. Roughly 56% of the world's ultimately recoverable coal resources are said to be in the USSR and Eastern Europe, 20% (including some very thick and shallow seams) in the USA, 9% in the People's Republic of China, 8% in Canada, 5% in Western Europe, 1.4% in Africa, 0.8% in Oceania, and 0.18% in Central and South America[7]. Much of the more accessible coal will cost tens of percent more than today's coal because of higher transport cost and the capital cost of new railways and pipelines; and not all of the cheap coal has conveniently low levels of sulphur and ash. World resources of solid hydrocarbons recoverable at costs probably within a factor of 2-3 of today's are sufficient, in the absence of climatic constraints (*infra*), to meet projected world demand for of the order of 200-300 yr[7], though this lifetime would drop to about a century if, as seems likely, large amounts of coal were converted to synthetic fluid fuels to make up for oil and gas shortages. (The complexities of *e.g.* coal needs for primary metallurgy, sulphur and ash content, and methods of transporting coal need not concern us on such a time-scale.)

[7] Hubbert, M K, "Energy Resources", in National Academy of Sciences / National Research Council, *Resources and Man,* W H Freeman (San Francisco), 1969.

Geopolitical constraints will probably remain less severe for coal than for oil.

Side-effects. The bulk of cheap coal and lignite reserves can be extracted only by surface mining on a very large scale[8]. Long-term restorative technology, still in a rather primitive state in most countries[9], may not be able to prevent eventual dereliction of fragile areas, though it should succeed in many others. Steep or arid land probably cannot be restored; elsewhere, great diligence and rigorous enforcement may be necessary[10]. Growing resentment at past despoliation and at such side-effects as acid drainage may be expected to make large-scale coal stripping politically awkward in such areas as the western USA. The social costs of coal mining, especially underground, are also high and will remain substantial for decades, but could be much reduced[11]: *e.g.* a recently opened Illinois deep mine is said to use every kind of safety and health precaution known, yet is still profitable. Highly or fully automated underground mining can probably be made as cheap as traditional methods, and, despite its engineering difficulty, must be a high research priority in many industrial countries, both for social reasons and to make narrow, wet, or steep seams recoverable. *In situ* techniques also may be needed for undersea coal, including the large North Sea measures.

[8] This generalization must be heavily qualified if "cheap" is broadly construed. If we assess resources ultimately recoverable at any price, then most of the coal and lignite is probably deep. In the USA, for example, only about 1/4 of reserves and 1/20 of resources is strippable—still, of course, an enormous amount of coal.

[9] Britain and Germany seem the most advanced. British costs for restoration to Grade B farmland often run about $1.40/m^2$ plus overheads.

[10] Box, T W *et al,* National Academy of Sciences, *Rehabilitation Potential of Western Coal Lands* (Ballinger for The Energy Policy Project, Cambridge, Mass., 1974); Gillette, R, *Science 182:*456 (1973).

[11] See *e.g.* Conti, J V, *Wall St J,* "Safety Underground. . .", 18 January 1973; reprinted in Committee on Interior and Insular Affairs, US Senate, *Hearings* Serial 93-7 (92-42), Pt 1, p 266, 22-3 March 1973 (USGPO, 1973); Dials, G E and Moore, E C, *Environment 16, 7, 18* (1974).

On the whole, it seems likely that an extensive cleanup of coal technologies, embracing both mining and combustion[12], would add only tens of percent to present costs. Devising such steps to minimize social and environmental costs of coal is an especially urgent task because in many industrial countries—including the USA, UK, and USSR—coal is the main bridge from the oil/gas to the energy-income economy. The responsible and orderly expansion of coal production is subject to unknown rate constraints, but they are probably not formidable. Such expansion will be much easier if we deploy coal technologies far more sophisticated than those now widely used; it is a *sine qua non* for a sustainable energy economy in many countries, and would be a more efficient and long-lived use of capital than most petroleum investments.

Fuel policy. The timing of shifts from petroleum to coal is controlled partly by important private-sector market forces too subtle to explore here. It is noteworthy, for example, that nearly all of the viable elements of the US coal and uranium industries, part of the nuclear component industry, most of the initially exploitable oil-shale and tar-sand reserves, and most of the known US geothermal resources are already controlled by the oil industry—which will soon control the new coal conversion industry and is reportedly buying many of the private US solar R&D companies too. Throughout the world, corporate integra-

[12] A potentially limiting hazard in coal combustion, especially with coal of high ash content, is the release of large numbers of sub-micron particles (1 micron = 1 μ = 10^{-6}m): Friedlander, S K, *Env Sci Tech 7:*1115 (1973). Such particles are very difficult to recover, and can be synergistic with other emissions (such as toxic metals) in their effects on public health; Natusch, D F S *et al, Science 183*:202 (1974). This problem deserves close study. The baroque approach now used for such awkward emissions as oxides of sulphur and nitrogen—passing them on to neighbours through a tall stack—is no longer good enough, as current studies of Scandinavian acid rainfall may show (see also Likens, G E and Bormann, F H, *Science 184:*1176 (1974)). It is all the less suitable for sub-micron particles, which can have very long atmospheric residence times. Another important area of ignorance is the distribution and effects of heavy metals naturally present in many fossil fuels.

tion is producing energy companies with strong monopolistic tendencies. This "diversification" is logical from the companies' point of view: they develop coal conversion because they have the money, and geothermal power because they have the drilling technology. The implications for a competitive marketplace, however, may be equally logical and rather less constructive[13]. It is not clear that the UK, with nationalized coal, is really better off, for the preferences of a monopoly consumer (such as the nationalized electricity industry) can and do turn the price structure of principal fuels upside-down.

A brief review of the proximate causes of present (pre-1973-war) US energy shortages may help to illustrate the failures, indeed the omissions, of fuel planning that are rapidly jeopardizing stable energy supplies all over the world. US errors of management have included:

a) Too-low natural-gas prices which destroyed "stretch" in productive capacity. (In consequence, prices must probably treble in this decade.)
b) Oil-import quotas which eliminated a strong reserve position: US oil production has already passed its peak, and the proposed shift from being the world's largest exporter to being the world's largest importer would be the most massive in present or future energy trade.
c) Stoppage of refinery construction for environmental and economic reasons; inadequate control over refinery operation.
d) Premature commitment of electric utilities to nuclear power.
e) Systematic neglect and running-down of the coal industry, even though on official projections US coal consumption is

[13] Somewhat different institutional pressures in the Federal Republic of Germany are now leading to the flooding of marginal coal mines—an effective way of burning bridges behind the nuclear power programme. Still another face of the same problem is the weakness of many national Governments in the face of multinational oil companies with considerably larger turnover than the GNP of all but the richest nations.

to treble by 2000. (Since the Oyster Creek contract[14] was signed in 1963, no new steam-coal mine has been opened anywhere in the USA without a prior long-term contract for selling the output.)

f) Heavy reliance of electric utilities on premium fuels.

g) Generally low energy prices which, coupled with the attitude that there will always be a surplus of cheap energy, has led to extravagant use.

h) Failure to anticipate and forestall the consequences of the above policies by appropriate research and development; *e.g.* sophisticated coal technologies must now be imported or improvised because there is almost no domestic technical base, most of the money having gone into development of fast breeder and other fission reactors. Coal gasification can therefore make no *large* impact before the late 1980s[15].

[14] This fixed-price contract (~ $117/kW(e) installed) for a nuclear power station was regarded as the first breakthrough into competition with coal. In the event, the station was commissioned two years late at a cost overrun of 32%. The utility's next order for the Oyster Creek site was postponed in favour of a coal-fired station.

[15] Synthesizing liquid and gaseous fuels from coal on a large scale will not be cheap or easy, though it can certainly be done at a cost—perhaps, for high-BTU gas, about 4-5× the present pipeline cost. The processes, particularly for high-BTU gas, are in preliminary or relatively advanced pilot development, with several methods being pursued at once. Construction of production plants could start in a few years if a high-cost, high-risk programme of parallel pilot and demonstration plants were launched now. Equally important is the development of much better technologies for burning coal: fluidized-bed combustion, a promising approach being studied in the UK and elsewhere, could be rather widely deployed in the 1980s*. See (on coal technologies in general) Hottel, H C and Howard, J B, *New Energy Technology—Some Facts and Assessments* (MIT Press, 1971) and Berkowitz, D A and Squires, A M, eds, *Power Generation and Environmental Change* (MIT Press, 1971). Squires gives an up-to-date review of gasification and fluidized-bed technologies in *Science 184:*340 (1974). Further useful surveys are by Osborn, E F, *Science 183:*477 (1974) and Nephew, E A, *Techn Rev* 76, 2, 20 (1973).

* Large-scale coal gasification is less than half as efficient as a less centralized system in which small fluidized-bed stations (perhaps based on gas turbines) produce hot water for local district heating plus electricity for remote heat pumps and other uses.

i) Failure to identify and regulate gross misuses of energy, including electric space heating; absolute rejection of energy conservation; preoccupation with increasing supplies rather than with first questioning demand; failure to regulate energy promotion.

j) Failure to maintain an adequate independent data base for future policy decisions.

k) Failure to evolve a suitable institutional base; about 64 Federal agencies now share jurisdiction over US energy policy[16].

Similar errors have occurred and are continuing to occur in the UK and most other industrial countries, where many of the decisions governing energy supply in the mid-1980s have already been taken, generally in a long-term policy vacuum.

Tar Sands

We have not so far referred to two possible sources of reduced carbon: tar sands, principally in Alberta, and oil shales, principally in Colorado and Wyoming, South America, and perhaps Asia[7]. The Athabascan tar sands of Alberta, in limited production since 1966, are now about to enter large-scale production for US markets. Recoverable world resources are approximately 1/5 to 1/7 those of liquid petroleum and are on average more (though not prohibitively) costly. Recovery by *in situ* retorting reportedly consumes about as much energy as is recovered (*cf.* a 5× direct-energy gain for surface mining); because about 90% of the Athabascan resource requires the former method, its net energy value may be very much less than its gross content. It seems unlikely that tar-sand products will be exported from North America; indeed, Canada is considering restrictions on exports to and investment from the USA.

[16] "Federal Energy Organization", Committee on Interior and Insular Affairs, US Senate, Serial 93-6 (92-41) (USGPO, 1973).

Oil Shales

Oil shale resources—under pilot study in the western USA for several decades—are potentially far larger than total world oil resources, but only if one incurs extreme costs by including submarginal shales so poor that the difficulties of production (including waste disposal) seem insuperable. One can now envisage the economic recovery of only about 1/10,000 of the total, equivalent to an order of magnitude less than world crude-oil resources. High-grade oil shales, which are about 10-20% hydrocarbons by weight, are difficult and expensive to process; are often under several hundred meters of overburden; occur mainly in arid regions where water supply[17] and runoff water quality severely constrain production; produce very large amounts of intractable waste rock; and yield an oil that may require special refining methods and equipment. The cost, waste, and water problems may be somewhat more tractable if large-scale *in situ* retorting can be developed, but capital and energy requirements will still be extremely large. There is no general agreement that oil shale will ever become an important energy source, and optimism would be decidedly premature, though limited production seems possible in the few regions possessing such reserves.

Comparison with coal. One leading authority[7] concludes that the shales "appear to be more promising as a source of raw materials for the chemical industry than as a major source of industrial energy." Compared with the latter course, production of low-sulphur liquid and gaseous fuels from coal seems cheaper, easier, and far nearer practicability[15, 18], despite its inherent

[17] For its requirement of perhaps 1-2 m^3 makeup water per barrel oil, an oil shale industry can be expected to compete very effectively with local irrigators.

[18] Coal conversion is several times as water-intensive as oil-shale conversion, but of the nearly 200 identified US sites with enough water and coal, some ought to be socially and environmentally suitable; and likewise abroad. The side-effects of coal conversion are speculative but

inefficiency, and can be expected to compete very effectively under any foreseeable price régime. (Material-handling problems are less with coal conversion than with shale because the energy yield per ton is about 4× as large.) There is also no doubt that world coal resources, though unevenly distributed, suffice[19] to support at least a century of intensive world industrial activity if the technical and political obstacles to their large-scale extraction and distribution can be overcome—obstacles which seem smaller than with any other centralized energy-supply technology of similar scale.

---

probably manageable, and the process could greatly reduce environmental and public-health impacts of combustion. *In situ* processes may conceivably simplify extraction and conversion of some deep reserves, through their side-effects are still to be evaluated. Some conversion schemes use a combined power cycle yielding both process heat (for the conversion) and electricity. The complexities of high- *vs.* low-BTU gasification need not concern us here, though the former, yielding gas of pipeline quality, would be more useful for basic shifts in energy supply over large regions.

[19] It must be noted, however, that large-scale consumption of coal is a medium-term measure only: fossil-fuel combustion in the long term will be limited not by coal resources but by climatic effects (see p 119).

5

Some Inner Limits

SOME INNER LIMITS

Rate and magnitude problems
Capital constraints

— 5 —

Rate and Magnitude Problems

The physical resource base does not tell the whole story of what can be done. It is hard to think of any thoughtful student of the fossil-fuel industries who believes that as world population doubles, say by the first decade of the next century, world energy conversion will increase 6×, as recent growth rates suggest. (This doubling assumes, for the sake of argument and prudence, that world population will in fact double, *i.e.* that food supply will do at least likewise. The author considers this very unlikely.) Even if the technical, environmental, political, and institutional difficulties of such an energy increase could be mastered, the rate and magnitude problems seem well beyond the capacity of world industry and finance—especially if they are to be devoted to other problems as well. In the author's view we should be hard pressed to produce half that much growth in the same period, particularly since world crude-oil production will undoubtedly peak within a generation. A mere doubling of world energy conversion would be hard enough: our earlier example of the limited effects of building one giant reactor per day would give the same results if we assumed any other energy technology.

Indeed, the exponential growth of energy conversion in any industrialized country (excepting such special cases as Norway) is the sum of a series of overlapping exponential curves representing new energy sources, each introduced as the preceding curve matures or begins to falter, and each in general of a *simpler technical character* than those preceding. Thus coal displaced wood, and was in turn displaced by oil and by gas (the latter, with its exceptional simplicity, accounting for 2/3 of US energy growth from 1945 to 1965). In most countries, domestic supplies of oil are in turn being supplemented by imported oil, and in some, gas by imported gas (LNG). Continued exponential growth of total energy conversion requires that *each successive source be capable of more rapid growth than that preceding it*—possible only if there is a very large existing energy inventory capable of being cheaply *and rapidly* extracted, processed, and distributed, as was historically true of Persian Gulf oil and US natural gas.

Governments have apparently acceded to quick depletion of cheap energy reserves on the tacit assumption that a new source will turn up in time to maintain ever faster growth. *The latest such innovation, however, is nowhere in sight,* and a little reflection about the character that such a new source must have reveals why. Nuclear power, with its complexity and long lead times, is far too slow. Existing energy industries are thus under severe stress to try to make up a growing deficit. (In the USA, for example, gas and oil fields are operating at 100%, or somewhat more, of their nominal maximum capacities.) This stress is in a sense due, as is often said, to various temporary shortages caused by prolonged mismanagement; but the underlying rate and magnitude problem is real and will not go away.

Capital Constraints

The rate at which energy technologies can be deployed is constrained not only by engineering logistics, but also by the rate at which investment capital can be generated or diverted. The rapidly rising capital intensity of the energy sector as nuclear fission, offshore drilling, LNG transport, coal conversion, etc are introduced entails capital demands so sudden and so large that rapid deployment soon makes them prohibitive: the capital-to-output ratio rises so quickly that companies accustomed to generating investment capital from revenues suddenly find themselves in the utility business, floating bond issues. Since Governments and large industries normally drive their fiscal machinery just within range of their investment headlights, estimating their total capital allocation only a year or two beyond the lead time of major energy projects, they cannot perceive in time that plans for rapid energy growth may demand unacceptable diversions from competing sectors or may even overcommit the total capital available[1].

The same is true on a world scale: *e.g.* it is not clear where the world oil industry will find $\$2 \times 10^{12}$ or more by the late 1980s, particularly in an era of runaway inflation and threats of global depression. But even in one specialized sector within one country, the point is often missed: *e.g.* the Central Electricity Generating Board's recently rejected 36-reactor ordering proposal for England and Wales in 1974-83 would have cost more than £11 billion, or 3× the CEGB's present net assets[2].

[1] In its May 1974 report, "US Energy Prospects: An Engineering Viewpoint" (also available as an "Executive Briefing Summary"), the National Academy of Engineering suggests that US energy independence in 1985 would require investment of $\$5\text{-}6 \times 10^{11}$. (In his 31 Oct 1973 remarks before the Financial Conference, National Coal Board, R C Holland of the Board of Governors of the Federal Reserve System estimated $\$7 \times 10^{11}$ and cited estimates of nearly $\$10^{12}$.) The NAE figure, averaged over the next decade, is at least half of the present rate of investment in all US industry and about twice the present rate of investment in the US energy sector. Even such sanguine estimates of attainable sustained rates of capital generation or diversic.. probably rest on cost estimates that are too low because they ignore the economic effects of shortages in materials and skilled labour (Policy Study Group, MIT Energy Laboratory, *Techn Rev* 76, 6, 22 [1974], pp 46-47). Moreover, most of the capital would be raised in public markets, now suffering severe loss of confidence (*Business Week,* 25 May 1974, p 102): the Consolidated Edison debacle is the first of many among US electric utilities, which are said to consume a quarter of all new capital generated in the USA, yet are suddenly curtailing their investment plans, especially for nuclear plants (Jacobs, S L, *Wall St J,* 19 July 1974, p 26) because their bonds have lost their attraction.

[2] This optimistically assumes for the unfuelled power stations a cost 20% less than a typical US turnkey price of about $700/kW(e), installed about 1982 as an 1150-MW(e) light-water reactor, assuming 2.30 $/£. See "Study of Base-Load Alternatives for the Northeast Utilities System", A D Little Inc, Cambridge, Mass., 5 July 1973.

The tendency of capital intensity in the energy sector to rise faster than energy output is normally considered only in the context of fiscal policy. In thermodynamic terms, however, its implications are more disquieting, for capital is ultimately energy: *money is not an independent entity, but must be referred back to real physical resources.* Increasing capital intensity of energy technologies means that *the energy cost of obtaining useful energy is increasing; hence that net energy yields from gross energy resources are declining*[3].

As the energy sector demands more capital, gross energy consumption and GNP could both in theory continue to rise while net energy available stayed the same or even fell. This is probably not happening yet, but without energy accountancy far more sophisticated than we have today[5], we cannot be sure. We also cannot assess how much *net* energy can be obtained from the resources which we list as *gross* and which thus seem larger than they are.

Moreover, energy technologies which now yield only a small energy profit will be even less energy-profitable, or even net consumers, as the high-grade (low-energy-cost) energy available to subsidize them declines. Odum remarks (*loc cit*[3], emphasis added): "This truth is often stated backwards in economists' concepts because there is inadequate recognition of external changes in energy *quality.* Often they propose that marginal energy sources will be economic later when the rich sources are gone. An energy source is not a source unless it is contributing yields, and *ability of marginal sources to yield goes down as the other sources of subsidy become poorer.*"

We accept the net energy losses of intensive agriculture[4] because we desire its unique product. Energy technologies

[3] This has been most forcefully stated in *Ambio 2:*220 (1973) by H T Odum, whose 1971 claim that we eat potatoes made not of soil but partly of oil was greeted at the time with some skepticism. Recent calculations by Leach[4] show that our potatoes are in fact made wholly of oil with oil left over.

[4] Leach, G, "The Energy Costs of Food Production", in Bourne, A, ed, *The Man-Food Equation,* Academic Press (London), 1974.

[5] See ref. 9, p 98 for surveys of various methods of net energy analysis.

running at a net energy loss are more insidious, for if unperceived they may seduce us into a commitment to pour our high-grade production-subsidizing energy ever faster down a rathole. Even on the most primitive analysis, such a possibility must today be taken seriously for *e.g.* oil-shale and fission technologies.

Many energy facilities now take several years' production to pay off their own energy cost. If heavy-engineering capital were worth an aggregated 1 GJ/$ (obviously a maximum boundary value, since it corresponds to spending *all* one's money on products refined from $5/barrel landed crude), then a 1-GW/(e) nuclear power station at current US costs, or $700/kW(e)—equivalent at 75% load factor to $63,000 per daily barrel of equivalent capacity—would require 42 years' cost-free output at 75% load factor to pay back the primary energy invested in it as capital. This 42 years would, at the other extreme, be only about 3½ if we more plausibly assumed the ~12× lower energy content of capital characteristic of aggregated UK GNP in recent years. The correct value may lie somewhere in between and can be found only by detailed energy analysis. It thus behooves us to examine far more carefully the gross-net distinction in capital-intensive energy technologies: *e.g.* the net energy yield of nuclear fission technology as a whole, taking *all* inputs into account, has not been shown to be positive, nor do the data exist on which such a calculation could be soundly based.* *(See note on following page.)*

It is interesting that at the conservative energy value of 10^8 J per dollar of capital investment[1] in major industrial facilities, investment of $\$6 \times 10^{11}$ corresponds to a direct energy cost of order 10^{10} barrels of oil equivalent—*just to build* the energy facilities required for Project Independence. (The actual energy cost could be twice this value.) In contrast, targets stated in the official "Project Independence Background Paper" (Washington Energy Conference, 11-12 February 1974) show that Project Independence would increase US oil and oil-shale production by amounts whose cumulative value, 1974-95, is 1.3×10^{10} barrels. Rapid deployment of energy technologies with long energy payouts and large initial energy investments can lead to an energy deficit on *current* account: in other words, a crash programme to increase energy supply can create an acute energy shortage!

A final and very important implication of the capital intensity of modern energy technologies is that even very expensive energy-conserving measures would be economic. For example, the Preliminary Report (*Exploring Energy Choices,* Ballinger (Cambridge, Mass.), 1974) of The Energy Policy Project suggests for the USA in the year 2000 a difference equivalent to about 33 million barrel/day between "historical growth" and "technical fix" scenarios. If that energy were to cost, say, $6/barrel and if a daily barrel of capacity were to cost $7000, then we could implement "technical fixes" costing over $200 billion initially plus $200 million/day—and still come out ahead! Furthermore, we would still have the fuel (but not the environmental and political side-effects of producing it). The actual costs of practically any sort of energy conservation are, of course, far lower than these figures. Thus we cannot afford *not* to conserve.*

* The best available data, together with a lucid discussion of net-energy dynamics, can be found in Price, J H, "Dynamic Energy Analysis and Nuclear Power", Friends of the Earth Limited (9 Poland St, London W1V3DG, England), 18 December 1974.

6

Nonfossil-Fuel Options

NONFOSSIL-FUEL OPTIONS

— 6 —

Energy technologies not based on fossil fuels offer options of widely varying time-scale and complexity. We shall now explore these options, bearing in mind the inner limits just described. It is here, in the realm of potential innovations, that the political, environmental, and social side-effects of new technologies need the most careful assessment, bearing in mind the very different needs and aspirations of different cultures. It is easy, for example, to fall into the trap of assuming that a particular technology is insignificant on a world scale because it can supply only, say, 1% of world energy needs. What matters is not aggregate energy production, but the ability of a technology to meet the energy needs of *people* in particular circumstances. A highly diffuse, low-intensity energy source might be ideal for a dispersed rural population. It would be imprudent for even the most highly urbanized and industrialized nations to assume that they will not or should not have substantial populations in this state.

Nuclear Fission

The promised innovation on which most energy planners have relied for the past two decades, apparently without considering how fast it can be deployed, has been nuclear fission—the extraction of heat from fissile (or, indirectly, from fertile) isotopes. This technology has received an enormous amount of highly skilled and dedicated attention for several decades. Yet Great Britain, the country now most heavily reliant on fission, generates only a tenth of her electricity in this way; and in the

Nuclear Fission and Fusion

Fission splits atoms apart; fusion sticks them together. Both release enormous amounts of energy. Fission is used in all functioning nuclear power plants (light-water, heavy-water, gas-cooled, and liquid-metal-cooled) and in atomic bombs. Fusion is used in hydrogen bombs and in the sun. The sun is 150 million kilometers (93 million miles) away: an admirable example of remote siting to protect energy users from the hazards of being too close to a nuclear source.

Fission is a sophisticated and exacting technology developed during the past few decades. In a fission reactor, the splitting of atoms releases energy which is collected as heat by circulating water, gas, or molten metal. The heat is then used to boil water, and the steam—as it does in a power station that burns fossil fuel—drives turbines whose motion is used to generate electricity. Meanwhile, the cooling fluid that boiled the water is cooled down and pumped through the reactor core again to keep the nuclear fuel from overheating and melting: fine tuning is imperative.

As a concentrated heat source for atomic power, scientists chose uranium-235, a dense and unstable metal found in small quantities in various parts of the world. By emitting tiny particles, uranium slowly changes into other elements, and eventually into the stable, dense, and well-known metal lead. But uranium-235 atoms, instead of decaying gradually, can split violently into fragments ("fission products") which are themselves radioactive. In the process, much energy is released, along with tiny particles called "neutrons". If the neutrons strike other uranium-235 atoms, they can cause these atoms in turn to fission. If sufficiently many uranium atoms are gathered together in a small enough space, a chain reaction begins, with each fission causing one or more further fissions. The tighter the configuration and the purer the uranium-235, the more explosive the reaction. The uranium in a light-water reactor is not pure enough to be capable of causing a nuclear explosion (though violent explosions might be caused instead by steam or chemical reactions).

Uranium atoms struck by neutrons do not always fission. Sometimes they change instead into other kinds of atoms—"transuranic" elements. Uranium-238, a very common kind of uranium, cannot itself be used to fuel a nuclear reactor, but can be changed, or "bred", into a manmade transuranic element called plutonium-239 that is usable as a reactor fuel (or as raw material for an atomic bomb). This breeding of nuclear fuel is the purpose of the "fast breeder reactors" now being developed in several rich countries. The uranium and plutonium in the core of a fast breeder reactor is in theory pure enough to cause a nuclear explosion if compressed during an accident; whether this might occur is a subject of debate. Still another nuclear fuel, uranium-233, can be made by bombarding thorium-232 atoms with neutrons. Uranium-233, like plutonium-239 and relatively pure uranium-235, can also be used to make atomic bombs. Heavy elements such as thorium, uranium, and the transuranic elements are together called "actinides"; they are radioactive and generally very poisonous. So are the fission products, especially strontium-90 (^{90}Sr), caesium-137 (^{137}Cs), iodine-131 (^{131}I), and the like.

Fusion, like fission, is a nuclear process, though the two have little in common. Fusion is touted as the answer to the problems of fission. Whether it can fulfill this role is uncertain, since a controlled fusion reaction has not yet been produced on earth, and decades of research and development remain to be done. Whether fusion (if feasible) will be economically and environmentally acceptable is still speculative.

Instead of splitting heavy, unstable metal atoms, fusion merges the nuclei of very light atoms under very high temperatures: for the simplest reaction (between deuterium and tritium, two rare kinds of hydrogen), a temperature of about 100 million °C is required. This might be achieved in several ways: in one, ultra-high-powered lasers would crush a hydrogen pellet to about 10,000 times normal liquid density, thus heating it to the required temperature. Heat could be carried away by helium gas or by other means. Several other possible methods do not use lasers. The main problem at the moment is finding materials or methods (such as strong magnetic fields) that can hold a hot enough and dense enough configuration of hydrogen nuclei together for long enough to complete the thermonuclear reaction, and yet withstand the stresses and radiations generated by the reaction. Lasers of great power, and giant magnetic devices for the several non-laser fusion techniques being experimentally studied, are now being developed. Of course, the energy applied to starting and containing the fusion reaction must be less than the energy obtained from it, or there would be no point in going to all that trouble.

USA, after 25 years and many billions of dollars' worth of research and development, nuclear power has only just surpassed firewood[1] as an energy source. Indeed, it was not until 1971 that the total annual electricity production of all US nuclear power stations exceeded the annual electricity consumption of the gaseous-diffusion plants used to enrich civilian and military uranium. In short, nuclear power has been very slow in arriving. Yet its advocates claim that it is poised for a remarkable spurt of growth with a doubling time of the order of a few years—sufficient to take over the role of now-faltering gas production in maintaining the recent world energy doubling time of about 12½ years. Most government and industrial planners share this view.

Limited but acrimonious public debate over whether this is possible or wise has exposed a deep division of opinion within the scientific community over fundamental technical and moral issues raised by nuclear fission[3]. Because this controversy is so crucial to any assessment of future energy supply, it is worth analyzing at some length. If the present summary seems critical, however, no reflection is intended on the motives of anyone engaged in promoting or developing nuclear fission. Decent, sincere people, perhaps caught up in the momentum and the intense social pressures of a committed organization, naturally like to think that what they are doing is worthwhile, and are not

[1] Firewood provided 3/4 of US energy in 1870, when US population was 1/5 today's and per-capita energy conversion[2] 2/7. Some observers believe the contribution of wood is underestimated and still exceeds that of fission.

[2] Freeman, S D, *Bull Atom Scient* 27, 8, 8 (1971).

[3] For one authoritative statement of "grave and justified misgivings", see the Statement from the Continuing Committee, 23d Pugwash Conference (Aulanko, Finland): *Pugwash Newsletter 11:*1+2 (July & October 1973). A fuller and more technical discussion of the facts and issues raised in this section appears in the author's paper "Nuclear Power: Technical Bases for Ethical Concern", submitted in autumn 1974 as evidence to the Royal Commission on Environmental Pollution, London, and available from Friends of the Earth Limited (9 Poland St, London W1V3DG, England).

always so concerned with gaps between intention and performance.

Issues. The problem of evaluating civilian fission technology has been well stated by Kneese[4]:

> "It is my belief that benefit-cost analysis cannot answer the most important policy questions associated with the desirability of developing a large-scale, fission-based economy. These questions are of a deep *ethical* character. Benefit-cost analysis certainly cannot solve such questions and may well obscure them. . . .The advantages of fission are much more readily quantified in the format of a benefit-cost analysis than are the associated hazards. Therefore there exists the danger that the benefits may seem more real than the hazards. We are speaking of hazards which may afflict humanity many generations hence, and distributional questions which can neither be neglected as inconsequential nor evaluated on any known theoretical or empirical basis. This means that technical people. . .cannot legitimately make the decision to generate such hazards on the basis of technical analysis."

Kneese is arguing, then, not merely that the benefits and risks of fission technology have been insufficiently quantified to support a benefit-cost judgment, but further that such a judgment could not be technically valid. Let us explore the reasoning behind this opinion.

The very large inventories of fission and activation products in the nuclear fuel cycle create risks[5] unlike those of any other

[4] Kneese, A V, "What Will Nuclear Power Really Cost?", *Not Man Apart 3,* 5, 16 (May 1973), Friends of the Earth, 529 Commercial St, San Francisco 94111; based on testimony in the USAEC hearings on the nuclear fuel cycle, November 1972. The full text is reprinted as "The Faustian Bargain" in *Resources* (Resources for the Future Inc, 1755 Massachusetts Avenue NW, Washington, DC 20036), September 1973.

[5] Ford, D F *et al., The Nuclear Fuel Cycle,* Union of Concerned Scientists (1208 Massachusetts Avenue, Cambridge, Massachusetts 02138), October 1973, reprinted Sept 1974 by Friends of the Earth Inc (529 Commercial St, San Francisco, California 94111).

single technology. These risks combine the geographic range of military pathogens, the permanence of irreversible changes in climate or in soil fertility, and the medical and moral significance of the most persistent synthetic mutagens. Because this unique combination of hazards departs so much from our experience, we must define with special care, before we choose to incur these hazards, the limits of our ability to cope with them.

Most fission technologists realize that they are creating new categories and magnitudes of risk, and respond with ingenious precautions. But by stressing the great care that they take, fission technologists evade the central question: are the safety problems of fission too difficult to solve? If they are, then (as Alfvén points out[6]) one cannot claim that they are solved by pointing to all the efforts made to solve them.

An engineering problem or a people problem? It is *impossible to prove,* except by experiment, whether or not the safety problems of widely proliferated fission technology are too difficult to solve. In assessing the risks of a complex technology in which "no acts of God can be permitted"[6], we can only rely on *analogies* with other highly engineered systems whose inherent risks are in principle several orders of magnitude smaller and quite different in kind. Such analogies suggest that safety in nuclear fission is ultimately limited not by our care or ingenuity (as has been true of all previous technologies[7]), but by our inescapable human fallibility: limited not by our ability to solve problems on paper, but by our ability to translate paper solutions into real events. If this view is right, as many distinguished scientists now believe,

[6] Alfvén, H, *Bull Atom Scient 28,* 5, 5 (1972).

[7] Save perhaps manned space flight. In *The New Yorker,* 11 and 18 November 1972, and in *Thirteen: The Flight That Failed* (Dial, New York, 1973), H S F Cooper Jnr gives a fine account of the impact of human fallibility on a highly engineered system (Apollo 13); *cf.* William Bryan's testimony on 1 February 1974, Subcommittee on State Energy Policy, Committee on Planning, Land Use, & Energy, California State Assembly. Kneese[4] also reminds us that "three astronauts were incinerated due to a very straightforward accident in an extremely high-technology operation where the utmost precautions were allegedly being taken."

then nuclear safety is not a mere engineering problem that can be solved by sufficient care, but rather a wholly new type of problem that can be solved only by infallible people. Infallible people are not now observable in the nuclear or any other industry.

International experience at many stages of the nuclear fuel cycle supports the thesis that describing nuclear safety problems as "amenable to engineering solution" confuses the way things are with the way one would like them to be. This experience shows that people have impressive talents in overcoming foolproof systems, and suggests that catastrophes have so far been averted more by luck than by design.

Fallibility and containment. Fallibility will find its greatest scope for expression in three exercises: containing radioisotopes within the nuclear fuel cycle, containing them after they have been rejected as wastes, and containing strategic materials. The third of these problems—perhaps the most critical—we shall discuss later, together with some risks of non-nuclear energy technologies; the other two are briefly discussed here. Such "fallibility problems" can be expected to become more prominent as reactors proliferate, salesmen outrun engineers, investment conquers caution, boredom replaces novelty, routine dulls commitment, and less-skilled technicians take over (especially in countries with "comparatively low levels of technological competence and a great propensity to take risks"[4]).

The degree of containment required against accidental or malicious release of radioactive inventories in the nuclear fuel cycle can be inferred from Holdren's calculation[8] that in a 1-GW(e) light-water reactor at equilibrium, 1/4 of the ^{131}I inventory would suffice to contaminate the atmosphere over the 48 coterminous United States to an altitude of 10,000 m to

[8] Holdren, J P, "Radioactive Pollution of the Environment by the Nuclear Fuel Cycle," 23d Pugwash Conference, Aulanko, 30 August—4 September 1973; revised and reprinted in *Bull Atom Scient 30,* 8, 14 (1974). (Dr Holdren is at the Energy and Resources Program, U. of California at Berkeley.)

twice the maximum permissible concentration (MPC) for that isotope; likewise, that half of the ^{90}Sr inventory would suffice to contaminate the annual freshwater runoff of the same area to 6× the MPC. Such numbers are not used to suggest that such dispersion would actually occur, but to stress that reliable containment deserves the most diligent attention.

Failure of containment seems most likely in three phases of the fuel cycle:

1) Reactor operation (where radioactive inventories are of the order of 10^{10} curie[9] and where the consequences of a major failure could include 10^4 or more deaths at ranges approaching 10^4—10^5 m, numerous injuries of widely varying induction time, and property damage of the order of \$$10^{10}$ over an area of perhaps 10^7—$10^{11} m^2$). Failure may be induced by disruptive energy releases (chemical or nuclear), mechanical or cooling faults, acts of war, or sabotage. The last two of these risks, the difficulty of decommissioning, and perhaps the magnitude of potential radioactive releases in any accident[10] could be much reduced by underground siting. This would cost about the same[11], but has

[9] One curie = 3.7×10^{10} disintegrations per second, roughly the activity of a gram of radium; limiting body burdens of the more dangerous radioisotopes are typically measured in millionths of a curie or less.

[10] It is often assumed, but may be untrue, that impermeable rock is a sufficient bar to the escape of volatile and semi-volatile fission and activation products after a core meltdown. Small or slow leakages are important; shafts must somehow be infallibly sealed; and contact with moisture would evolve prodigious quantities of steam, hydrogen, etc.

[11] Rogers, F C, *Bull Atom Scient 27,* 8, 38 (1971); Smernoff, B J, "Underground Siting of the LMFBR Demonstration Plant: A Serious Alternative", HI-1686/2-P, Hudson Institute (Croton-on-Hudson, New York), 12 Sept 1972; Aerospace Corp. ATR-72 (S7263)-1, March 1972 (Cal Tech Environmental Quality Laboratory Report no. 6, Sept 1972); ORNL SD-71-66-(R)-5872 (Oak Ridge National Laboratory, August 1966); Jamne, P, IAEA-A/CONF 49/P/290 (Geneva Conference, Sept 1971); Olds, F C, United Engineers & Constructors, Inc, UEC-UNP-710701, July 1971, and *Power Engineering,* Oct 1971, p 34; Blake, A *et al,* UCRL-51408, Lawrence Livermore Laboratory, May 1973.

attracted almost no official interest[12, 13] despite the likelihood that such fission plants, often sited near cities or major military targets, "will be enormously attractive objects for sabotage and blackmail"[14, 15]. Alternatively, risks can be greatly increased by raising power density, as in large light-water reactors[16] or in fast breeder reactors (~400MW/m^3)[17].

[12] Wilson, C L, "A Plan for Energy Independence", *Foreign Affairs,* July 1973.

[13] The USAEC recommended it to the President in 1962. The proposal sank without trace, presumably because awkward questions might be asked about reactors already built on the surface.

[14] Edsall, J T, *Science 178:*933 (1972), eloquently restated in *Environmental Conservation I,* 1, 32 (1974); *cf.* DeNike, L D, *Bull Atom Scient 30,* 2, 16 (1974).

[15] The large stored energy of reactors aids malicious intervention: the mechanical energy alone in the primary circuit of a large pressurized-water reactor is equivalent to $\sim 3 \times 10^4$ kg TNT.

[16] It is often claimed that the probability of major component failures in these reactors is extremely low; but we do not really know that, and the reactors appear not to know it either. The USAEC series *Reactor Operating Experiences* casts serious doubt on the claimed independence and predictability of complex events (Hanauer, S H, and Morris, P A, IAEA-A/CONF 49/P/040 at Vol III, p 205, Fourth [UN] International Conference on the Peaceful Uses of Atomic Energy (Geneva, Sept 1971), IAEA, Vienna, 1972) and on the assumed improbability of major failures. (See also the USAEC compilations ORNL-NSIC-17,-64,-69, -87,-91,-103,-106,-109, and -114.) Even if all our experience of all reactors were relevant to large reactors of high power density, it would not suffice to prove that catastrophic accidents from random causes are as unlikely as 10^{-3} per reactor-year. (The extent of operating experience is reviewed in Lovins and Patterson[39],especially in Annexes A and C.) It is hardly reassuring that practical defects or inadequacies in any of the three elements of the "defence-in-depth" philosophy are cheerfully shrugged off by reference both to the presence of the other two elements and to the alleged but unverified improbability that they will be needed—even though some types of failure in one element could *cause* failures in others; nor that the list of the "unresolved" generic safety issues named by the USAEC's Advisory Committee on Reactor Safeguards (*e.g.* in the ACRS letter of 13 February 1974) seems to be steadily expanding. The much-published Rasmussen report (USAEC

The disposition, during an accident, of a fast breeder's fissile inventory (including[18] ^{239}Pu equivalent to about 10^2 bare-sphere critical masses) cannot be predicted for a partially molten core affected by sodium flow or compression waves. It is hard to see how any convincing assurance against a Bethe-Tait accident (critical reassembly) can be given[19]. Superprompt criti-

---

WASH-1400), issued in draft in August 1974, calculates accident probabilities and consequences to be very low; but even the initial stages of independent review have revealed significant methodological flaws, and it seems likely that the Rasmussen conclusions (Gillette, R, *Science 185:*838 [1974]) will be placed in doubt by orders of magnitude: see Lovins, A B, *op. cit.*[3].

[17] High power density and high plutonium content mean shorter thermal and neutron-physics time constants, placing a heavy burden on control systems and on any negative coefficients of reactivity. Large stresses and high temperatures also aggravate metallurgical or geometric faults in materials whose long-term behaviour under extreme neutron flux is still mainly conjectural.

[18] Plutonium-239 is the most plentiful of the artificial radioisotopes made by bombarding uranium with neutrons. It is pyrophoric and reactive, is a hard alpha emitter with a 24,400-year half-life, is a bone-surface-seeking poison similar to radium but several times as toxic, readily forms respirable aerosols, and is one of the most hazardous substances known. See Tamplin, A R and Cochran, T B, "Radiation Standards for Hot Particles" (Natural Resources Defense Council, 1710 N St NW, Washington DC, 20036, 14 February 1974). The plutonium in breeder reactor fuel is already oxidized into a refractory ceramic which, it is claimed, cannot produce respirable particles; but in an accident, this claim may be incorrect. (It is known that oxide fuel can be at least partially reduced by hot sodium.)

[19] Elaborate computer calculations have been made of the course of disassembly after assumed reassembly accidents. Basic physical understanding of the course of reassembly, however, is inadequate for the task and has progressed scarcely at all in the past two decades. *Cf.* McCarthy, W J and Okrent, D, "Fast Reactor Kinetics", in Thompson, T J and Beckerly, J G, *The Technology of Nuclear Reactor Safety,* Vol. I (MIT Press, 1964/70); Lovins, A B, *New Scientist 61:*693 (14 March 1974); Cochran, T B, *op cit*[25]; Jackson, J and Boudreau, J E, "Postburst Analysis (Preliminary Report)", internal document, Argonne National Laboratory, December 1972; Kelber, C N *et al,* ANL-7657, Feb 1970.

cality resulting in a substantial nuclear explosion is probably much less likely than a low-yield disassembly—a nuclear fizzle—sufficient to breach containment and release the transuranic and fission-product inventory. There is virtually no limit to the size of the catastrophe that this mechanism could produce.

2) Transport of irradiated fuel. It is impossible to design a shipping cask that can withstand all accidents, let alone sabotage. Risks—not calculable in detail but apparently underestimated—will increase with proliferation, especially with a ^{239}Pu economy and associated shipment of short-cooled fuel. The US envisages 100 railway shipments daily; a single fast-breeder fuel discharge would contain 5×10^8 curie of activity and emit 1.7 MW of heat.

3) Reprocessing of irradiated fuel. Criticality accidents excepted, internal energy release is less likely here than in a reactor; but the containment is generally less, and the radioactive inventories and concentrations are vastly larger. A more lethal target for acts of war or acts of God can hardly be imagined: a major release could make large regions permanently uninhabitable.

The 1-to-3-ton ^{239}Pu inventory in a fast breeder reactor poses an especially formidable containment problem because of its extreme toxicity. A lethal dose for everyone on earth could probably be contained in a piece the size of an orange; indeed, orders of magnitude less would probably suffice[18] (a piece the size of a marble, for example, if Tamplin and Cochran are correct). World inventories are planned to rise rapidly from a few tens of tons now to hundreds in the early 1980s and thousands several decades hence.

Some radioisotopes mobilized by the nuclear fuel cycle, *e.g.* ^{226}Ra and its decay products (released in uranium mining and milling), deserve but seldom receive proper containment[5].

High-level wastes emerging from the nuclear fuel cycle can be converted from corrosive solutions to stable solids with technologies now in hand. This rational precaution, if applied to liquid wastes as soon as they are produced, could substantially reduce the risks of temporary storage, but it brings us no closer to discovering a method of permanent disposal—of perpetual[20] and infallible isolation from the biosphere. The volume of highly active material requiring such disposal is still relatively small but is growing rapidly (to of order 10^{12} curie by 2000), and its period of significant potency—of order 10^3 years for fission products[21] and 10^6—10^8 years for transuranic activation products[22]—subjects any terrestrial disposal scheme to geological requirements of which we have no experience and for which no responsible geologist can offer a guarantee. The time-scales involved are less the province of geology than of theology. Terrestrial disposal, *e.g.* in basement rock or in salt beds, must therefore be "retrievable"—and hence requires "surveillance" on a time-scale far exceeding the observed life-span of human cultures. Since safe extraterrestrial disposal is neither technically nor economically feasible, one could properly conclude that as a

[20] *E.g.* 10^6 kg of irradiated light-water-reactor fuel yields 50 kg of ^{241}Am, whose hazard (taking daughter products into account) "does not decrease significantly in more than ten million years": Isaacson, R E and Brownell, L E, at p 955 in OECD-NEA/IAEA, *Management of Radioactive Wastes from Fuel Reprocessing,* OECD 66 73 02 3, Paris, 1973, a valuable summary.

[21] Some minor fission products such as ^{129}I and ^{99}Tc last far longer.

[22] Kubo, A S and Rose, D J (*Science 182:*1205 [1973]) propose the thorough separation of these two categories so that the actinides can be burned in a reactor. The practical advantages of such a scheme may be substantial, though they are perhaps not as large as is claimed. For example, it would be interesting to know how much additional waste, including highly dilute alpha waste, is generated by the repeated recycling of extracted actinides. Moreover, reactor operation (and very large associated actinide inventories) would have to continue for a time long compared with the half-life of ^{239}Pu. The four-volume USAEC report *Advanced Waste Management Studies, High-Level Radioactive Waste Disposal Alternatives* (Schneider, K J and Platt, A M, eds, BNWL-1900, May 1974) does not adequately explore such implications of waste partitioning, nor indeed of other concepts.

technical problem, high-level waste disposal is probably insoluble even in large amounts of money, and that as a moral problem, it should deeply concern us. (Disposal of low- and medium-level radioactive wastes—solid, liquid, and gaseous—is perhaps a less serious problem, but is still very important, especially in long-term cumulative impact, and is seldom dealt with properly.)

Fission economies. It is instructive to contrast with these considerations official plans for several thousand large power reactors to be operating three decades hence. Further into the future, two prominent US energy planners[23] envisage 4,000 seaside fission "parks", each producing 40 GW(e) and nearly twice that much heat, and all producing a total of 7×10^{11} curie/yr of long-lived activity. The fuel would be about 1.8×10^{8} kg/yr of ^{238}U and ^{232}Th, derived from about 5.5×10^{12} kg/yr of granite (about twice the present world consumption of coal) and converted respectively to ^{239}Pu and ^{233}U. This scheme represents a 50× increase in total human energy conversion. The accompanying commitment of humanity "to exercise great vigilance and the highest levels of quality control, continuously and *indefinitely*"[4] has led some to propose a self-perpetuating "technological priesthood"[24]. Others reply[14] that the required long-term social stability "has not existed in the past, does not exist now, and offers no promise of existing in the near future"; that "even devoted priests are human"; and that the importance of the "priesthood" proposal resides in the support it lends to the view that the technical and social problems of the fission economy have no realistic solutions.

Though the economics of such a 50× energy increase depend on, and are as speculative as, those of the fast breeder

[23] Weinberg, A M and Hammond, R P, *Am Scient* 58:412 (1970).

[24] Weinberg, A M, *Science* 177:27 (1972). Dr Weinberg's recent speeches suggest that he is no longer so enthusiastic about this proposal: see *e.g.* "Can Man Live With Fission—A Prospectus", Woodrow Wilson Center for Scholars, Washington DC, 18 June 1973.

reactor now under development[25], it is important to note, in view of the special hazards of fast-breeder technology[25] and of the ^{239}Pu economy[26], that projected populations of thermal (non-breeder) reactors could be economically fueled for at least a century unaided by any breeder reactors at all[27]. This is because the cost of uranium feedstock (U_3O_8) is only a few percent of the sent-out cost of nuclear electricity[28]; hence the latter is insensitive to the former. A 2.2 $/kg increase in the price of U_3O_8 would increase the marginal cost of bulk electricity from a 1-GW(e) pressurized-water reactor[29] by only 0.052 m$/kW-h without, or 0.034 m$/kW-h with, Pu recycle[30]; the cost increase would be less for the consumer, who also pays fixed

[25] Lovins, A B, in "Hearings on the LMFBR Demonstration Plant", Joint Committee on Atomic Energy, US Congress, 8 September 1972, pp 120-130, excerpted *Bull Atom Scient 29,* 3, 29 (1973), a polemical survey; Lovins, A B, *New Scientist 61*:693 (14 March 1974), a précis; Cochran, T B, *The Liquid-Metal Fast Breeder Reactor: An Economic and Environmental Critique* (Johns Hopkins University Press / Resources for the Future, 1974), a detailed technical assessment; Gillette, R, *Science 182:*38 (1973), *184:*877 (1974); Hammond. A L, *Science 182:*1236 (1973), *185:*768 (1974). The Natural Resources Defense Council's comprehensive comments on the AEC's draft of its breeder programme Environmental Impact Statement (WASH-1535) were released in May 1974 and update Cochran's book. They are to be reprinted in the final draft of WASH-1535, probably in 1975, but are currently out of print.

[26] Very large ^{239}Pu inventories would accumulate in the nuclear fuel cycle notwithstanding that a full ^{239}Pu economy would take a very long time to achieve (owing to the long doubling times likely to be attainable and to the large ^{239}Pu deficits during early decades of breeder programmes).

[27] Holdren, J P, "Uranium Availability and the Breeder Decision", Environmental Quality Laboratory Memorandum #8 (Cal Tech, Pasadena, 1974). *Cf.* Rose, D J, *Science 184:*351 (1974), and Bupp, I C and Derian, J-C, *Techn Rev 76,* 8, 26 (July/Aug 1974).

[28] Larson, C E, "International Economic Implications of the Nuclear Fuel Cycle": S-11-72 (given 10 July 1972), AEC *News Releases 3,* 32, 3 (1972).

[29] US Atomic Energy Commission, WASH-1139 (Rev 1), December 1971, p 10.

[30] Larson, C E (Commissioner, USAEC), personal communication, 2 November 1972.

distribution costs, and should be similar for gas-cooled reactors. Thus a 10× increase in U_3O_8 price—enough to increase *e.g.* US reserves by more than 2000%[31] without requiring environmentally unacceptable mining methods[32], and similarly in many other countries[33]—would increase the bulk price of present nuclear electricity by about 18-27%, trivial compared with increases in other energy sectors[34]. This is not to say that either thermal or fast reactors are a good idea, but only that proponents of the latter over the former have not made a valid case[25].

Sources of delay. Public awareness of the issues raised by nuclear fission is now sufficient to make fast proliferation of this technology difficult in many countries despite its immense political, economic, and institutional momentum. Fuelled both by deep and rapidly growing anxiety in the technical community

[31] US Atomic Energy Commission, "US January 1, 1971 High Cost Uranium Resources—$30 to $500/lb U_3O_8", unpublished sheet dated "9/1/71". It appears from geochemical arguments and drilling statistics that the AEC's estimates of intermediate-grade U resources, as in this sheet, are far—perhaps 10× or more—too low[27]. Unlike many rare metals, U appears to have essentially Laskian distribution statistics down to very low grades.

[32] Bienewski, C L *et al*, "Availability of Uranium at Various Prices From Resources in the United States", US Bureau of Mines IC-8501, 1971, describes Chattanooga Shale recovery which would consume large land areas and which might yield little or no net energy via thermal reactors.

[33] Uranium, and for that matter thorium, is surprisingly widely distributed. Many of the richer deposits are in North America, Southern Africa, and Australia, but a prominent French official has recently estimated that France alone could supply the EEC's U_3O_8 needs for well over a century at about $88/kg. Many countries probably have large intermediate-grade conventional U and Th resources. Even U recovery from seawater might not be wholly crazy if large amounts of seawater are being pumped through power stations for cooling.

[34] Cost comparisons between fission and fossil-fuel power are meaningless anyhow because of strong government "incentives" aiding mainly the former: Alfvén, H, *Bull Atom Scient 30,* 1, 4, (1974). (As David Lilienthal remarks, incentives cost the same as subsidies, but sound nicer.)

(especially in the USA[35]) and by the obvious need for any reëvaluation to begin at once before the commitment becomes irreversible, public debate will quickly expand and intensify. Rapid shifts in opinion are illustrated by recent actions of official bodies in California, Sweden, the Federal Republic of Germany, and the United Kingdom[36].

Perhaps surprisingly, however, the much-publicized delays in US reactor commissioning in the past few years, and the failure of presently planned reactors to meet more than half of the early-1980s trend projection, have been due almost entirely[37] to poor management, manufacturing and labour difficulties, poor quality assurance[38], unexpected technical snags, and the

[35] Warnings have come from such sources as the RAND Corporation ("California's Electricity Quandary", R-1116-NSF/CSA, September 1972), the Federation of American Scientists (*FAS Newsletter,* February 1973), former top officials of the USAEC, and scientists and engineers of stature. On 26 October 1973, the Council of the American Physical Society backed its distinguished working party, which had concluded, after briefings by top AEC and other safety experts, that the APS should promptly conduct an independent technical study of reactor safety.

[36] Select Committee on Science and Technology, House of Commons, "The Choice of a Reactor System" (First Report 1973-74) (HMSO, London, 1974); see also §55 and Annexe B in Appendix 1 to the Minutes of Evidence[39]. On 10 July 1974, the UK Government accepted the Select Committee's advice and rejected a proposal to order light-water reactors. A letter in the *Financial Times* (8 June 1974) by the eminent metallurgist Sir Alan Cottrell, formerly Chief Scientific Advisor to the same Government, also appears to have been influential.

[37] Doub, W O, speech to Atomic Industrial Forum, San Francisco, 12 November 1973: S-13-73, AEC *News Releases 4,* 47, 13 (1973). For some reasons why interventions are ineffectual, see Eblin, S and Kasper, R, *Citizen Groups and the Nuclear Power Controversy,* MIT Press, 1974.

[38] RAND[35] cite "increasing reports of poor quality control and documented carelessness in manufacture, operation, and maintenance"; many USAEC reports tell the same story, *e.g.* Morris, P A and Engelken R H, IAEA-SM-169/47 (Jülich, February 1973) in *Principles and Standards of Reactor Safety,* IAEA (Vienna), 1973. The original (October 1973) version of the AEC Task Force Report "Study of the Reactor Licensing Process" reports for the period 1 January 1972—31 May 1973 approximately 850 "abnormal occurrences [which] involved malfunc-

other standard problems of rapid high-technology proliferation at rapidly rising size and power density—not to public resistance or intervention in licensing proceedings (despite very serious doubts[39] about the operating safety of US light-water reactors). It is these former conventional constraints, the rate and magnitude problems we stressed earlier, that will prove decisive in preventing fission from meeting projected short- and medium-term demand even where public opposition is absent, suppressed, or ignored. Nuclear power does not represent a large reservoir of cheap energy capable of being mobilized very simply and quickly; it is on the contrary one of the most complex and unforgiving technologies known to man. Some people still think that nuclear capacity in, say, the USA will increase 50× by 2000 and that the equivalent of total present US electrical

---

tions or deficiencies associated with safety related equipment. . . . Many of the incidents had broad generic applicability and potentially significant consequences. . . .[This] raises a serious question regarding the current review and inspection practices both on the part of the nuclear industry and the AEC."

[39] The depth and basis of these doubts—particularly doubts about the effectiveness, reliability, and relevance of emergency core cooling systems (ECCS)—are seldom realized by those unfamiliar at first-hand with the extensive public record of the Bethesda ECCS rulemaking hearing (USAEC RM-50-1). This record shows that most of the AEC's top safety experts believe, contrary to the official AEC position, that there is insufficient technical basis for establishing adequate margins of safety (and hence for licensing the reactors). For an annotated review of LWR safety and economics, see Lovins, A B and Patterson, W C, Appendix 1 (pp 145-178), Appendices to the Minutes of Evidence (73-vii), "The Choice of a Reactor System", Select Committee on Science & Technology (Energy Resources Sub-Committee), House of Commons (HMSO, London, 1974). A more technical safety review, summarized in part in *Environment 14,* 7, 2 (1972), is the detailed and strongly recommended Concluding Statement by the Union of Concerned Scientists (1208 Massachusetts Avenue, Cambridge, Massachusetts 02138), reprinted in 1974 by Friends of the Earth Inc (529 Commercial St, San Francisco, California 94111). For political background, see also Gillette, R, *Science 177:*771, 867, 970, 1080 (1-22 September 1972).

capacity will then be built every 29 months. There is no accounting for what some people think.

Nuclear Fusion

Prospects. There is perennial speculation, and much misunderstanding, about a nuclear technology whose scientific feasibility, unlike that of fission, remains to be proved (probably in the next decade). Nuclear fusion could be practically free of resource constraints, since the ^2H (deuterium) easily extractable from seawater[40] represents about $5 \times 10^5 \times$ the energy content of all fossil fuels. (Lithium reserves adequate for a few centuries could also provide an easier thermonuclear reaction.) It is this feature of fusion that has led to many years of research in several technologically advanced countries. It is likely, though not certain, that at least one of several promising approaches[41] will succeed. This could lead to a commercial prototype around the last decade of this century, at a cost of the order of several billion dollars, followed by the usual rate and magnitude problems of proliferation. Such a device (whose economics are completely conjectural) will probably be extremely complex, and it seems implausible that the needed technology would be made freely available to all Governments (as the fuel would).

Side-effects. The safety problems of a practical fusion reactor cannot be assessed without building one—a course of action that can be argued to be prudent if feasible. Sources of disruptive energy release may include strong magnetic fields and large

[40] It is generally assumed that traces of deuterium in seawater have no biological function; nobody really knows.

[41] Magnetic confinement of plasmas is now within about an order of magnitude of the goal. No fundamentally new mistake has been made for several years; on the contrary, some 175 types of plasma instabilities have now been classified. Laser implosion of pellets (Emmett, J L *et al, Sci Amer 230,* 6, 24 [1974]) is conceptually simpler but poses awkward materials problems. It may also have military implications.

inventories of molten lithium (comparable to the molten sodium in a fast breeder). The reactor could not get out of control, nor create strategic materials[42]. The inventories of radioactive products would be significant (of order 10^8 curie of ^3H) but could in theory be recycled as fuel. The main problem would be activation products produced by the reactor's intense neutron flux, comparable to that in the core of a fast breeder. The theoretical biological significance of these products[43] may on present estimates[44] be an order of magnitude smaller (though considerably shorter-lived) than in a comparable fission reactor. Unless it is many orders of magnitude smaller than that—which is possible but not yet plausible[45]—fusion does not represent an attractive long-term energy source, though in the medium term it is almost certainly preferable to fission. One could argue further that if fusion turned out to be a clean, cheap, and safe energy source[46], man would lack the discipline so to manage it as to prevent serious ecological and climatic damage (*infra*). Recent history does not say much for man's responsibility in managing large stocks of energy. This argument suggests that

[42] Tritium and other light nuclei could perhaps be a significant exception when the design principles of thermonuclear weapons eventually leak out. Neutrons from a fusion reactor or from a sub-thermonuclear reactor (in which *e.g.* an energetic ^3H beam is fired into a ^2H plasma) could also breed fissile isotopes.

[43] The difference between theoretical and actual containment could be as important for fusion as for fission: fallibility and economic expedience could compromise safety analogously in both cases.

[44] Post, R F and Ribe, F L, LA-UR 73-1684, Los Alamos Scientific Laboratory, 1973.

[45] This depends mainly on how wide the choice of materials for the containment structures turns out to be[44]. Alternatively, if extremely high temperatures can be attained, it may be possible to use certain fusion reactions which release almost no neutrons; this would be far more difficult than the reactions presently contemplated, but would virtually eliminate activation products.

[46] Some would say that no nuclear source can be clean and safe; and that though nuclear power is admirable when properly sited, the source and user should be rather widely separated—say, about 150 million km.

fusion is not a panacea and may bring more problems than it solves: thus, that we should not try to develop fusion power unless we are quite sure that we shall have the technical options and the political will to reject it if we should. Otherwise a coal economy would be the best bridge to an economy of energy income during the first half of the next century.

Geothermal Heat

Another possible energy source which could be considered consumption of capital rather than of energy income[47] is geothermal heat, now drawn only locally in the form of hot water and dry steam. Some research, of which there has been far too little so far, suggests that geothermal energy from hot dry rock, if obtainable, might be cheap and abundant enough to have a profound influence on world energy supplies. Side-effects are still speculative, but could well be more acceptable than those of fossil fuels, despite low thermal efficiency. Little more can be said of this source[48] until further research is completed a few years hence. It would take a large geothermal capacity, though, to yield as much low- and medium-temperature heat as electrical generating stations now waste—$\sim 3 \times 10^{11}$ W in the USA alone.

[47] The uncertainty here is semantic, not technical. The outward flux of geothermal heat is continuous, but a heat-collecting scheme would generally deplete local stocks of heat faster than they are replenished, thus in effect consuming temporarily a slowly renewable capital stock.

[48] Cornell Workshop on Energy and the Environment, *Summary Report,* Committee on Interior and Insular Affairs, US Senate, Serial 92-23, May 1972 (USGPO, 1972); Rex, R W, *Bull Atom Scient 27,* 8, 52 (1971); "Geothermal Resources and Research", Committee on Interior and Insular Affairs, US Senate, Serial 92-31, 15 and 22 June 1972 (USGPO, 1972); Robson, G R, *Science 184:*371 (1974); "Energy from Geothermal Resources", Committee on Science and Astronautics, US House of Representatives, Serial Q, May 1974 (USGPO, 1974).

Energy Income

ENERGY INCOME

Tides
Hydroelectricity
Indirect solar collection
Direct solar collection

— 7 —

Having exhausted the known categories of energy capital usable in large amounts, we can now consider possible technologies for exploiting energy income—energy flows which natural processes make continuously available whether we tap them or not. Such techniques, suitably deployed, are perfectly capable of providing as much energy as mankind could reasonably need. Along with geothermal, fusion, and advanced coal technologies, energy-income technologies have suffered greatly from several decades' overwhelming stress on fission R&D; Alfvén suggests[1] that we "must imagine how [the 'unconventional']. . .sources of energy would appear today if research and development had been concentrated on them."

Tides

Tides, the only such source not activated by the sun, can provide locally large blocks of physically clean (if ecologically disruptive) power from especially favourable sites, sometimes at competitive prices. For fundamental physical reasons, however, they cannot provide more than about a hundredth of the power potentially available from conventional hydroelectric stations. For this reason, and because of their special site requirements and relatively high capital cost, they are likely to remain of merely local interest.

Hydroelectricity

Hydroelectricity, the largest indirect source of solar energy now used for generating industrial energy, could in theory be

[1] Alfvén, H, *Bull Atom Scient 30,* 1, 4 (1974).

expanded more than tenfold to provide several times the total present world electrical capacity. Such full use will not be approached in practice because:

a) The accumulation of silt renders most manmade reservoirs useless within a century or two (or even less); this problem may well be insoluble.

b) As recent controversies have shown in the USA, where only about 1/5 of theoretical capacity has been used, land-use conflicts can delay or prevent large hydroelectric developments.

c) Near- or even over-saturation of prime sites in most industrialized countries means that the great bulk of unused theoretical capacity is in poorer countries unable to build their own dams and sometimes unable to obtain enough foreign help without onerous conditions. (South America, Africa, and Southeast Asia contain respectively 20, 27, and 16% of the world's theoretical capacity[2].)

d) Recent experience with large dams, especially in tropical countries, suggests that the geophysical, ecological, epidemiological, and social side-effects of such massive projects may make them a very bad bargain. Smaller, more labour-intensive projects and carefully integrated land-use planning might significantly reduce such side-effects, but in the present state of knowledge it would be premature to reach wholly sanguine conclusions.

e) Large water projects could have regional and even global climatic effects[3]—especially projects affecting the Arctic, *e.g.* diversion of Siberian rivers. Such effects cannot be accurately forecast with present knowledge but are the cause of considerable concern. A combination of constraints (d) and (e) probably makes most major projects (*e.g.* Mekong, Amazon) inadvisable.

[2] Hubbert, M K, "Energy Resources", in NAS/NRC, *Resources and Man,* W H Freeman (San Francisco), 1969.

[3] Study of Man's Impact on Climate (SMIC), *Inadvertent Climate Modification,* MIT Press (Cambridge, Massachusetts), 1971.

It seems unlikely that installed world hydroelectric capacity will double during this century. If it did, the increment would equal several percent of present world energy conversion. The dependence of many poor countries on their hydroelectric resources, however, does make it very important for them that the above-mentioned problems be vigorously attacked.

Indirect Solar Collection

Solar energy can be collected indirectly by the following methods:

a) Agriculture, forestry, fish-farming, hunting, fishing, and gathering, all of which are beyond the scope of this paper. Such sources embody immense design experience and, since they make their solar collectors from the energy flow which they process, require no fossil-fuel subsidy. Properly run farms can be nearly self-sufficient in energy even if mechanized[4].

b) Synthetic agricultural systems designed to produce fuel by photosynthesis. This looks less promising (for reasons of efficiency and ecological stability) than the use of natural productivity (general biomass, or agricultural and human wastes) through direct combustion, distillation, or fermentation. Like combustion of urban solid wastes, these techniques, now used locally on a small scale, could be quickly developed as a useful supplement to conventional energy sources in many regions. Much more work is needed in this field: routes to methanol and methane (*infra*) offer attractive major energy options.

c) Windmills are not an ideal source for centralized energy

[4] Leach, G, "The Energy Costs of Food Production", in Bourne, A, ed, *The Man-Food Equation,* Academic Press (London), 1974. *Cf.* Pimentel, D *et al, Science 182:*443 (1973), and Slesser, M, *J Sci Fd Agric 24:*1193 (1973).

conversion but can be significant[5]—not only in rural settlements—and should not be ignored. Their efficiency has greatly improved in recent years. The related problem of energy storage is briefly considered below; for wind-power on a moderate scale it is not difficult.

d) Though harnessing of sea currents does not seem practicable, recent suggestions of operating heat engines from vertical temperature gradients in tropical oceans[6] may merit further study. Costs and side-effects are unknown; large-scale use in the next few decades seems rather unlikely. The technology could be explored through bottoming cycles in electricity generation.

Direct Solar Collection

We have saved for last the possibility of directly collecting the diffuse[7] but enormous solar flux to generate high-temperature heat or electricity, for this is a major medium- and long-term option which, after decades of persistent but nominal effort, is just starting to receive the attention it deserves. (The recent Japanese commitment to major efforts in solar research and development—now being echoed in the USA—is a straw in the wind.) Centralized conversion to electricity on a very large scale must, however, be distinguished[8] from diffuse conversion, *e.g.*

[5] Heronemus, W E, "The US Energy Crisis: Some Proposed Gentle Solutions", ASME/IEEE paper, 12 January 1972; from author, Dept. of Civil Engineering, U. Mass., Amherst, Mass. 01002. See also his "Pollution-Free Energy from Offshore Winds", 8th Annual Conference and Exposition, Marine Technology Society, 11-13 Sept. 1972, Washington, D.C. (A Feb 1974 doctoral thesis suggests circumstances in which the stated cost could be reduced 2× in a larger offshore system.) See also Salter, S H, *Nature 249:*720 (1974) on wave-power.

[6] Zener, C, *Physics Today* (January 1973), p 48.

[7] A conservative round number for average insolation at *high* temperate latitudes is 100W/m^2.

[8] "Solar Energy Research" (staff report), Committee on Science and

in single-dwelling collectors that yield hot water and (via heat pumps[9]) hot and cold air. The latter type of device is in use in many countries today, albeit in rudimentary form, and would probably be competitive with conventional methods in most temperate latitudes if supported by modest development efforts. Indeed, there is good reason to believe[8] that diffuse solar technology may already be competitive with conventional methods anywhere in the USA or similar latitudes. There seems no fundamental reason why such domestic-utility technologies could not be widely proliferated within the next few decades, producing vast savings of fossil fuels. The barriers to this innovation appear to be entirely institutional—inertia, lack of military applications, lack of incentives for fuel vendors, inappropriate economics, etc. It is not generally realized that modern "selective black" surfaces—highly absorbent through the visible spectrum but poor radiators in the far infrared—can attain very high working temperatures even on a cloudy day. After all, the effective working temperature of the sun as an energy source is about 5500°K—nearly the boiling point of tungsten, and much hotter than a boiler.

Centralized photothermal or photoelectric conversion is a somewhat peculiar concept, since one of sunlight's greatest virtues is free distribution. Major engineering studies of large

---

Astronautics, US House of Representatives, Serial Z, December 1972 (USGPO, 1973); Donovan, P *et al,* "An Assessment of Solar Energy as a National Energy Resource", NSF/NASA Solar Energy Panel, University of Maryland, NSF/RA/N-73 001, December 1972 (PB-221 659, National Technical Information Service, Springfield, Virginia 22151); Wolf, W, *Science 184:*382 (1974); Morrow, W E Jnr, *Techn Rev 76,* 2, 31 (1973).

[9] Properly engineered domestic heat pumps would be very useful anyhow. Kovach, E G, ed, *Technology of Efficient Energy Utilization,* NATO Science Committee Conference (Les Arcs, 8-12 October 1973), Scientific Affairs Division, NATO (Brussels, 1974). *Cf.* Berg's comments on the industrial sector in *Science 184:*264 (1974).

centralized systems are still embryonic, but suggest[10] that capital costs are likely to be within striking range of present sources if reduced by research and development aimed at developing better materials[8]. Thin-film sandwiches and Fresnel-lens arrays look particularly interesting, as do some organic photovoltaic systems. Low maintenance cost, low environmental impact, free fuel, and security of supply should weigh heavily against any residual increment of capital cost. It is argued that photothermal conversion requires a great deal of land, but in fact it requires less than prolonged strip-mining of coal, and can co-exist with certain other uses. The total energy budget of large-scale solar collectors needs careful study, lest they yield less net energy than photosynthetic/chemical-conversion routes.

Solar power in general has several unique implications which do not arise from its obvious advantages (such as reliability, simplicity, guaranteed safety and feasibility, sufficient magnitude, free fuel, and low environmental impact). For example, it could help to redress the severe energy imbalance between temperate and tropical zones; its diffuseness is a spur to decentralization and increased self-sufficiency of population (highly desirable on other grounds); as the least sophisticated major energy technology, it could greatly reduce world tensions resulting from uneven distribution of fuels and from limited transfer of technology; and by limiting the density and the absolute amount of power at man's disposal, it would also limit the amount of ecological mischief he could do, without necessarily limiting his ability to live a comfortable and happy life. The climatic impact of a direct or indirect solar economy, finally, would be minimal.

[10] Ford, N C and Kane, J W, *Bull Atom Scient 27,* 8, 27 (1971); Meinel, A B and M P, *Bull Atom Scient 27,* 8, 32 (1971); Meinel, A B and M P, "Briefings Before the Task Force on Energy", Vol. III, Committee on Science and Astronautics, US House of Representatives, Serial U, pp 32-52, 6 March 1972 (USGPO, 1972) — *cf* testimony in 5 June 1973 hearing No 12 before the same Committee (USGPO, 1973). A L Hammond describes in *Science 184:*1359 (1974) a typical advance in current materials research.

These special advantages would not be shared by extraterrestrial solar collection, as in Glaser's proposal for large photoelectric collectors in synchronous orbit. The economics of such a scheme are several orders of magnitude out of reach, though they may deserve reëxamination several decades hence. The safety problems of transmitting the power to earth via focussed microwave beams may be soluble. Possible military applications of such a microwave source may account for official interest in the scheme.

Energy Systems And Components

ENERGY SYSTEMS AND COMPONENTS

— 8 —

Completeness requires, but brevity does not permit, a detailed discussion of energy conversion, storage, and distribution technologies, which may have more impact on energy supply than does the availability of primary fuels. A few short conclusions may give some of the flavour of recent work in this complex field:

a) If approximately 2/3 of the heat produced by primary fuel in the generation of electricity continues to be wasted, electrification should be strongly discouraged as inherently inefficient: indeed, most average efficiencies have *fallen* in the past few years. Correcting the scandalous waste of fairly-high-temperature heat from power stations, as with the district-heating schemes common in Sweden, could reduce this objection gradually over the next few decades[1]. "Thermal pollution", in other words, could be put to good use which would otherwise demand fresh primary energy. Meanwhile, the electric industry plans to use 1/2 of US primary energy in 2000 (now 1/4), which would require 2/3 of the *total* continental runoff[2] as cooling water to take the "waste" heat.

[1] The importance of using "waste" heat for heating or industrial processes can hardly be over-emphasized. The obstacles are less technical or economic than institutional. To reduce these it may sometimes be worthwhile to degrade high-temperature heat deliberately in order to insert a stage of buffering between the power station and the users. Surprisingly, large heat pumps may be a superior source of process heat. Kovach, E G, ed, *Technology of Efficient Energy Utilization,* Scientific Affairs Division, NATO (Brussels, 1974); Berg, C A, *Science 184:*264 (1974).

[2] Singer, S F, "Environmental Quality and the Economics of Cooling", in Berkowitz, D A and Squires, A M, eds, *Power Generation and Environmental Change* (MIT Press, 1971).

b) Magnetohydrodynamics (MHD) now looks less promising than combined-cycle gas turbines as a way of improving electrical generating efficiency. Early versions of the latter already are about as efficient as the best modern steam-raising plant, and promise dramatic further improvement. Cryogenic generators and transmission lines also hold much in store.

c) Electricity can be stored more flexibly, efficiently, and cheaply in underground compressed-air reservoirs than by hydroelectric pumping—which is already uneconomic compared with peaking gas turbines. Battery technology continues to improve but no real breakthrough is yet in sight. Electricity is still probably the most expensive form of energy to transport in bulk.

d) Recent experiences of electrical utilities strongly suggest that the economies of scale alleged for large central stations and components may be illusory. Not only is component reliability subject to important diseconomies of scale, but the whole subject of centralized *vs* diffuse energy conversion needs fundamental restudy.

e) Heat storage is not as difficult as is often supposed, especially on a small scale. Large-scale schemes for underground storage of hot water[3] may be a relatively easy answer and deserve study.

f) Chemical energy storage—as refined petroleum products, natural gas in gaseous or liquid form, hydrogen, alcohols, etc—is an immensely intricate subject linked with *e.g.* transport engineering, fuel-cell technology, and industrial chemistry. Research can probably enable us to assess on paper the implications of the theoretically elegant and flexible hydrogen economy[4] and of the possible advantages

[3] Meyer, C F and Todd, D K, *Envir Sci Techn* 7:512 (1973). Some other energy storage options are reviewed by Robinson, A L, *Science 184:*785, 884 (1974).

[4] Several recent informal studies suggest that a hydrogen economy, properly designed, need not be unduly hazardous. More work is needed. *Cf.* Cahn, W R, *Nature 248:*628 (1974). The National Technical

of a safer, probably cheaper, and more easily adopted methanol economy[5]. Methanol is already cheaper in long-haul transport than LNG[6], and offers both a smooth route of transition from an oil to a hydrogen economy and a convenient way of storing hydrogen in *e.g.* vehicles—a carbon and an oxygen atom serve, in effect, as carriers for four hydrogen atoms.

g) Microbiological and enzymatic methods of energy conversion, ranging from the Calvin/Tributsch photoelectric cell to astonishingly efficient methods of producing hydrogen, methane, and methanol from simple substrates and then reconverting these fuels in biological fuel cells, may produce many surprises in the coming decade[7].

Obviously, no "unconventional" energy technology, or combination of technologies, offers an instant panacea; and the utility of particular choices will depend on local needs, which vary as widely as the range of fossil-fuel options. One can, however, argue that any strategy of research and development should favour (and foster by incentives) as diverse as possible a range of options. Needs change, technological adventures fail, plans need constant revision; but in diversity lies safety. The possible virtues of decentralized energy conversion need special stress: they may include lower cost, greater flexibility, smaller

---

Information Service (P O Box 1553, Springfield, Virginia 22151) publishes *Hydrogen Energy,* a 1953-73 bibliography with summaries and a quarterly update service. A typical technical study of part of the hydrogen economy is Wentorf, R H Jnr and Hanneman, R E, *Science 185:*311 (1974).

[5] Reed, T B and Lerner, R M, *Science 182:*1299 (1973); Davis, J C, *Chem Eng,* 25 June 1973, p 48.

[6] Dutkiewicz, B, "Methanol Competitive with LNG on Long Haul", 52d Annual Meeting, Natural Gas Producers Association, Dallas, 26-28 March 1973; *Oil & Gas J* (30 April 1973), p 166. Methanol plants have recently been ordered in place of LNG facilities.

[7] C-G Hedén (Karolinska Institute, Stockholm) has surveyed such options. Melvin Calvin's fine summary of photosynthetic routes (*Science 184:*375 [1974]) suggests many intriguing possibilities.

strategic vulnerability, higher system efficiency, lower risk, greater resistance to monopoly, compatibility with a wider range of human settlement patterns, and (depending on the technology considered) greater self-sufficiency in a world of increasing interdependence and instability. Energy independence is already becoming a major aim of public policy in those countries whose indigenous resources (ranging from coal to low latitude) permit it; and current political events will surely speed the trend.

Those energy technologies that would be most useful to the largest number of people in the world are simple, require sound engineering but no glamorous new technologies, and are now receiving little attention. They would perform basic functions—cooking, lighting, heating, and pumping—and would be based partly on cheap wind- and light-collectors (*e.g.* selective-black plates) and partly on a simple, foolproof device for converting a wide range of organic materials to methane or methanol. Such energy technologies are undoubtedly the ones the world needs most, and would have a rapid and profound effect not only in the rural villages of poor countries, but in all countries that have enough common-sense not to be mesmerized by more flashy and elaborate devices.

9

Energy Conservation

ENERGY CONSERVATION

Forecasting
Doing more with less
Social patterns and goals

— 9 —

Most of this paper has been concerned with what an economist would call "energy sources" rather than "energy uses". While trying to redress this imbalance in the short space remaining, we must not make the common error of supposing that "energy demand" is any less important than "energy supply". Just the opposite is true today even though the emphasis of policy and strength of advocacy have long favored demand overwhelmingly over supply. In the next few years we shall learn that however much energy people "demand", they cannot use it if it is not available. People will learn to conserve energy whether they want to or not. From now on the choice will become rapidly clearer: whether to reconcile demand with need according to an orderly plan, or by panicky improvisation in the face of imminent shortages. The nations most responsible for these shortages can reflect that were their people not using several times as much energy as they need, cheap energy supplies could be available to the rest of the world for some decades to come. If per-capita energy use in the USA were reduced to that of, say, France, the amount "saved" would suffice to give everyone else in the world nearly a fourth more energy than he now has. Likewise, it is said that 200-odd million Americans use more electricity for air conditioning than 800-odd million Chinese use for everything!

A few simple examples, easily multiplied, will illustrate the scope for energy conservation[1,2,3] in the industrialized nations.

[1] Perry, H, "Conservation of Energy", Committee on Interior and Insular Affairs, US Senate, Serial 92-18 (USGPO 5270-01602), 1972.

[2] Kovach, E G, ed, *Technology of Efficient Energy Utilization,* NATO Science Committee Conference (Les Arcs, 8-12 October 1973), Scientific Affairs Division, NATO (Brussels, 1974); Berg, C A, *Science 184:*264 (1974); Over, J A, Sjoerdsma, A C, Eds. "Energy Conservation: Ways and Means", publication 19, Stichting Toekomstbeeld der Techniek (Prinsessegracht 23, the Hague), 12 June 1974.

[3] Hannon, B, *Techn Rev* 76, 4, 24, (1974); Berg, C A, *Ibid,* p 15.

a) Most modern architecture is extremely energy-intensive, with absurdly high lighting levels, poor insulation, universal air conditioning, electric heating, excessive volume, excessive use of energy-intensive materials such as cement and aluminium, waste of locally available low-temperature heat, and non-solar orientation. The modern attitude in this sector is perhaps best exemplified by the World Trade Center in New York, a single building wired for 80 MW; by office buildings where neither air conditioners nor lights can be turned off, though if it gets too cold one can turn on the electric heater as well; by the way we heat our cities with power stations to run air conditioners that make our cities hotter; and by the apparent inability of the British and other Governments to prescribe for houses the minimal standard of roof insulation required for piggeries (though heavier insulation still, as is common in Scandinavia, would pay for itself in a year or two and would be perhaps the most cost-effective way of helping the UK balance of payments: in Sweden, even *triple* glazing of windows is now economical!).

b) In most factories that do not depend on exceptionally energy-intensive processes, energy savings of 10-20% at little or no gross capital cost can be quickly identified by consultants. The National Industrial Fuel Efficiency Service in the UK says that British industry could save about $1 billion per year in energy costs (at mid-1973 prices).

c) The extravagance of modern private transport[4] and the poor deployment of public transport are too obvious to need much comment. Transport accounts directly for about 25% of total US energy conversion, almost twice as much as for space heating and cooling. The load factor of cars—a sensitive leverage point—has been dropping so fast in the USA that William Ruckelshaus projects that "by 1980 one out of every three cars will be tooling around without a driver." Institutional reforms are badly needed in such areas as car insurance regulations and airline franchising (*e.g.*

[4] Rice, R A, *Techn Rev* 76, 4, 45 (1974).

perhaps commercial aircraft could be shared in a common pool, just as US railway freight hauliers share their rolling stock). Using canals, and even such unconventional methods as containerized airships, could be worthwhile for intercity freight transport: and sailing-ships have worked well for a long time. Better communication, too, could help us to move information rather than people, at far lower energy cost; and transport hardware is a good place to start pressing for durability and low-entropy design.

d) The energy cost of non-recycled materials[5] (about 1/7 of US energy) and of intensive agriculture (especially in the use of protein feedstocks) is unknown but undoubtedly sensational. A can may cost twice as much energy as the pound of vegetables inside contains[6]. The direct energy cost of 1 kg of protein in English milk is estimated to be equivalent to about 20 liters of petrol. A ton of gutted fish landed by a UK trawler costs a ton of oil. Slesser[7] cites a Stanford Research Institute estimate that modern US production of 1 kg of beef protein requires about 350 MJ, equivalent to about 12 kg of coal. Such energy subsidies are used to develop "chickens that are little more than standing egg machines, cows that are mainly udders on four stalks, and plants with so few protective and survival mechanisms that they are immediately eliminated when the power-rich management of man is withdrawn."[8]

e) Energy is substituted for political and moral as well as for physical resources. Preparations for war (the ultimate form of conspicuous consumption) account for several percent of world energy conversion. The potential energy represented by present stockpiles of nuclear weapons would, were it food, feed everyone on earth for a decade. The potential energy represented by US munitions detonated over Indo-

[5] Berry, R S and Makino, H, *Techn Rev* 76, 4, 33 (1974).

[6] Leach, G, "The Energy Costs of Food Production", in Bourne, A, ed, *The Man-Food Equation,* Academic Press (London), 1974.

[7] Slesser, M, *New Scientist* 60: 328 (1 November 1973).

[8] Odum, H T, *Environment, Power, and Society* (Wiley-Interscience, New York/London, 1971).

china, 1965-71, is equivalent to the food input of some two million people over the same period.

Energy waste cannot be avoided nor rational energy policies devised unless the energy costs of each technology and of each material flow are known. The development of detailed energy-added accounts is just the beginning[9]. Better information on where our enormous energy conversion goes and what we get for it, presented in sufficient detail that we can energy-budget *e.g.* fresh peas *vs* frozen peas (together with capital, labour, and other trade-offs), is likely to produce many surprises. It will encourage the evolution of a whole-systems approach that uses integrated design to match various forms of energy to the most appropriate uses and to minimize total conversion. This is likely to be cheapest in the long run and to minimize ecological damage.

In this connexion the intimate relationship of energy strategy to practically every other major area of policy must be stressed once more: there is no "energy problem" in isolation. We have already touched on the energy-intensiveness of modern agriculture. Still more complex is the role of energy in human settlements. The modern suburban agglomeration—especially the utopian slurbia—may have a power density so high, and so vulnerably expressed, as to be unsustainable in an era of limited and uncertain energy supplies. Plans for suburbanization must take this into account.

Forecasting

Energy forecasting, now in considerable ferment, needs to be rebuilt around causality rather than correlation, so that the

[9] The best survey papers on energy analysis are those prepared by M Slesser (University of Strathclyde, Glasgow, Scotland) for the International Federation of Institutes for Advanced Study (Nobel House, Sturegatan 14, Box 5344, S-102 46 Stockholm, Sweden) and P F Chapman's annotated article in the journal *Energy Policy* (IPC, London, June 1974). With many colleagues and under IFIAS auspices, the author is engaged in attempts to devise widely agreeable ground-rules for a new science of energy impact analysis.

validity of assumptions and the state of fundamental determinants of energy demand can be constantly reviewed. At present we do not know what these determinants are, nor what the price elasticity of demand might be for any particular form of energy, nor whether large price rises would have much effect on industrial economies as a whole (as opposed to *e.g.* the private transport sector; fuel costs in the former have in the past been a small fraction of total costs). A mature science of energy forecasting would also include an institution to anticipate and identify, and to assess the energetic consequences of, changes in technology or in market patterns[10]. Rational regulation of undesirable trends is impossible unless potentially energy-intensive products or practices can be noticed early, before decision-making becomes politically painful.

Doing More With Less

Though little is known of the social and physical role of energy in various societies, we now know enough to be confident that energy conversion in most rich countries is several times their actual need. We must act now to put this knowledge into practice by stages—trimming the energy fat (typically 30-40% of present levels) that can be trimmed without changing lifestyles substantially, while we decide on longer-term changes in direction.

Merely slowing energy growth[11], rather than slowly swinging it slightly negative, seldom buys the time needed for technical innovations on the required scale. In the USA, a politically acceptable strategy could be devised and implemented

[10] In its "full-steam-ahead" scenario, The Energy Policy Project[12] has had to assume a modest saturation in US energy demanded by 2000: not because supply constraints were assumed, but because the Project staff could not devise plausible ways to use that much energy.

[11] Wilson, C L, "A Plan for Energy Independence", *Foreign Affairs,* July 1973.

within two years for an orderly reduction of any residual growth rate by perhaps half a percentage point per year to a negative 1-2%/yr. The author believes that without prohibitive cost or disruption, and with considerable advantages, the total level of US energy conversion could then be reduced by a factor of at least two over the ensuing two or three decades. It is unlikely that anyone who has seriously studied the scope for energy conservation and for redeployment of economic activity in the USA will quarrel with this conclusion. Similar, though perhaps less drastic, economies are undoubtedly possible throughout the industrialized world, and are often necessary on other grounds (*e.g.* food supply). No world or national energy strategy can succeed without close attention to such measures. As Malcolm MacEwen remarks, a man who cannot fill his bathtub because the water keeps running out does not need a bigger water-heater; he needs a plug.

It may be argued that a civilization in, say, Denmark using only half as much electricity as now is inconceivable; but one existed in the mid-1960s when the Danes were at least half as civilized as now. What would the lifestyles of the mid-1960s look like now with more efficient use and more equitable distribution? What would the USA look like with a material standard vaguely similar to that of 1910 but much better distributed and applied more efficiently to more useful ends, and with such important but energy-cheap amenities as modern medicine and telecommunications? Such questions must be asked and answered now.

Social Patterns and Goals

The range of social options available to rich countries is far wider than a narrow trumpet-shaped future curving steeply upwards; futures based on energy stability or shrinkage are equally realistic in technical, economic, and social terms, as the important work of The Energy Policy Project[12] has just revealed. (It appears, for example, that the widely assumed dependence of a healthy economy and full employment on energy growth may not exist.) But democratic change requires credible options to be presented in detail and discussed in public. People cannot choose options they do not know about. And though present energy shortages are disruptive because of ad-hocracy and outmoded values—not through any inherent impracticability of thinking twice as hard and using half as much—there is a danger that people may be persuaded by energy vendors that a three-day week in Britain, going without hot water in Stockholm, etc is a foretaste of life in a rationally planned low-energy economy, rather than of life in an increasingly vulnerable high-energy economy.

The political implications of energy futures are as important as they are unperceived. A low-energy society might (but need not) be normative and pluralistic; but a high-energy society would probably be far less attractive to those who value freedom of personal choice. In the latter society, we would (as now) extrapolate the shape of the industrial economy, then derive from it our settlement patterns and their social consequences; but in the former, we might design diverse settlements where people with varying tastes would want to live, then devise an economy to support them. Such an approach might yield *e.g.* less *unwanted* mobility—"the multiplication of trips made necessary by the existence of vehicles"[13]—recognizing that "beyond a certain speed, motorized vehicles create new distances

[12] The EPP final report *A Time to Choose: America's Energy Future* (Ballinger (Cambridge, Mass.), October 1974) and many important specialized commissioned studies are essential reading. A list is available from EPP, 1755 Massachusetts Avenue NW, Washington, DC 20036. An EPP-commissioned econometric study of energy, employment, and economic growth is summarized in part by S B Sheppard on pp 69-70 of the 1 June 1974 *Business Week.* EPP-like exercises in many countries are to be conducted as part of the 1974-6 Workshop on Alternative Energy Strategies, 1 Amherst St, MIT, Cambridge, Mass. 02139.

[13] Illich, I, *Energy and Equity* (Calder & Boyars Ltd, 18 Brewer St, London W1R 4AS; 1974). *Cf. Tools for Conviviality, ibid,* 1973.

that they alone can shrink; they create them for all, but they can shrink them only for a few."[14] The philosophical departure in such an approach is profound, as Illich's refreshingly radical argument[13] on the costs of mobility suggests:

> "By now, people work a substantial part of every day to earn the money without which they could not even get to work. . . .The typical American male devotes more than 1600 hours a year to his car. He sits in it while it goes and while it stands idling. He parks it and searches for it. He earns the money to put down on it and to meet the monthly instalments. He works to pay for petrol, tools, insurance, taxes and tickets. He spends four of his sixteen waking hours on the road or gathering his resources for it. . . .The model American puts in 1600 hours to get 7500 miles: less than five miles an hour. In countries deprived of a transportation industry, people manage to do the same, walking wherever they want to go, and they allocate only three to eight percent of their society's time budget to traffic instead of 28 percent."

The social values that fuel a high-energy society are all too apparent today. Those that could make a low-energy society succeed are deeply rooted in many cultures, but may need to be rescued from societal attics and dusted off. What happened to thrift? to neighbourliness? to craftsmanship? to the notion that esteem is merited more by conspicuous simplicity than by conspicuous consumption? Recycling such values could help us to achieve (as Daly puts it) growth in things that really count, rather than in things that are merely countable.

To assess and choose paths for the orderly reduction of energy conversion in rich countries, we shall need detailed scenarios of various lifestyles at different levels of energy intensity: The Energy Policy Project of the Ford Foundation, a pioneering US effort[12] in this direction, will be deservedly influential, and its extensions to Denmark, the UK, and a wide

[14] Illich, I, quoted by Toynbee, T, *Observer* magazine, 24 February 1974, p 29.

variety of other countries will be instructive. We shall also need fuller details of the social, ethical, political, and humanistic implications of technical and social choice—more important, perhaps, than the technical and economic implications which, though important, do not deserve to monopolize the attention of decision-makers. It will take many years' hard work to develop all these tools; we cannot wait that long. But though we do not yet know exactly what to do, we have a good idea of what *not* to do, and we need not wait for every last fact to come in before we stop doing it. Large public projects centred round the private car—our biggest consumer ephemeral—offer an excellent example of resources that deserve a better fate.

10

Energy Issues

ENERGY ISSUES

Distribution
Risk
Ethics
Conflict

— 10 —

So far this paper has tried to give an overview more of facts than of issues. It is time now to review and stress some important issues that must affect our choice of energy strategies.

Distribution

First is the drastically skewed distribution of energy resources and of energy conversion. Correcting the latter will require far more responsive and far-seeing management in the industrialized countries, ready to formulate deliberate strategies for energy conservation[1] before rationing and failures of supply enforce a haphazard and spartan self-sufficiency[2]. It will also require societies with quite different goals to learn from the mistakes into which an expanding economy of flow has led their materially richer, though not always happier, neighbours—now beset by sectional efforts to secure a larger share of a pie that has stopped swelling. Furthermore, skewed energy distributions not only in

[1] The collected publications of The Energy Policy Project constitute the first serious attack on this problem; the MIT Workshop on Alternative Energy Strategies (*supra*) is to be the second. Both rest on the thesis that energy growth can be decoupled from growth in social welfare.

[2] The thesis that a lower-energy society must be unpleasant to live in does not withstand thoughtful analysis—indeed, the EPP "Zero Energy Growth" scenerio envisages substantial energy growth! Moreover, the policy tools required to encourage significant energy conservation are far less coercive than those required to maintain historical growth. This is correct even for lower-energy paths than those studied by EPP—*e.g.* for that suggested above in the section "Doing More With Less".

space but in time, as we steal subsidies from the future[3], raise issues fundamental to the morality of modern industrial societies[4].

Risk

Next is the equally important issue of risk. Rapid growth in energy conversion increases the magnitude and likelihood of mistakes. It also seems recently to have come to depend on the rapid development and deployment of technologies so complex and so novel that their side-effects may be unpredictable—or, more commonly, may be predicted but not taken properly into account in decision-making, owing to institutional momentum or to the distractions of a crisis atmosphere. It is instructive to study three outstanding examples of exceptional and little-known risks in new energy technologies:

a) Unirradiated or reprocessed fissile isotopes, with half-lives of 10^4 years or more, must be perpetually safeguarded against "diversion" (theft) by terrorist groups, non-nuclear Governments, lunatics, criminal syndicates, or speculators. A few kg of ^{233}U, ^{239}Pu, or highly enriched ^{235}U—regardless of chemical form—can be made into a crude but wholly convincing nuclear weapon[5] by a physicist with a

[3] As Hugh Nash remarks, what better way is there to show our faith in our descendents' boundless technological ingenuity than to make sure they need it?

[4] The problem of a sustainable energy economy is inseparable from that of a sustainable economy—an economy of stock in which growth is no longer used as a substitute for distribution (the "let them eat growth" principle) and in which moral growth is combined with a biophysical steady state. Daly, H E, *Toward a Steady-State Economy,* W H Freeman (San Francisco), 1973.

[5] The suitability of an isotope for weapons depends on *inter alia* the ratio of its fast-neutron fission cross-section to its non-fission capture cross-section. This ratio is rather favourable for some isotopes which are not fissile in the usual sense, *viz.* fissionable by *slow* neutrons. Study of the

small group of dedicated technicians, a $\$10^4$ model shop, less than a year, and techniques thoroughly described in the open literature[6]—mainly in publications issued by principal public nuclear agencies and thus likely to be considered authoritative. Under some circumstances, the resources required would be far smaller: a single fanatic could probably produce an effective bomb from certain materials in a few weeks[6]. Experience of fabricating strategic materials is available from thousands of people now in civilian life, but is not essential. Indeed, actual manufacture of a weapon is optional if a credible threat of having done so can be manufactured instead[6, 7]. Reactor-grade, rather than weapons-grade, plutonium is suitable for sophisticated amateur weapon-makers if an unpredictable but perhaps large yield is sufficient[8]: no effective denaturant for fissile Pu is known.

Thermal reactors produce only slightly less net ^{239}Pu

---

published cross-sections and neutron yields at the appropriate energies shows that a fast critical assembly can be made from reasonable quantities of certain isotopes which are not fissile, and some of which are *not* now safeguarded. The present discussion therefore refers to "strategic material", not just to the three fissile isotopes named or to "special nuclear material" as defined by USAEC regulations. We neglect here possible future developments, worthy of grave concern, connected with laser or centrifugal methods of isotopic enrichment.

[6] McPhee, J, "Profiles: The Curve of Binding Energy", *The New Yorker,* 3/10/17 December 1973, republished as *The Curve of Binding Energy* (Farrar, Straus, & Giroux, New York, 1974); Willrich, M and Taylor, T B, *Nuclear Theft: Risks and Safeguards,* Ballinger (Cambridge, Mass.) for The Energy Policy Project, 1974.

[7] Lapp, R E, "The Ultimate Blackmail", *NY Times* magazine, 4 February 1973.

[8] Gilinsky, V, *Environment 14,* 7, 10 (September 1972); Hall, D B, Los Alamos DC-13114 (1971); Mark, J C, "Nuclear Weapons Technology", in Feld, B T *et al,* eds, *Impact of New Technologies on the Arms Race,* MIT Press, 1971; Jauho, P and Virtamo, J, "The Effect of Peaceful Use of Atomic Energy upon Nuclear Proliferation", Helsinki Arms Control Seminar (Finnish Academy of Arts & Letters), June 1973.

than do breeder reactors[9]; hence safeguards problems are here and now. A 1GW(e) light-water reactor makes several hundred kg of ^{239}Pu a year. Highly enriched uranium fuel, such as is used in many high-temperature gas-cooled reactors and research reactors, is easier to fabricate, though less reactive. A homemade weapon which failed to achieve high yield could still do immense damage through prompt radiation[6] — and, in disassembling, disseminate ^{239}Pu and the like. Dissemination could also arise through malice or through incompetent attempts at theft or at weapons fabrication.

Extensive, mainly secret, and certainly inadequate measures[6,10] are now taken to prevent theft of strategic material. The safeguards systems literature shows that continual small thefts from fuel facilities are in principle undetectable; that it may take weeks or months to detect large single thefts; and that the precision of inventory assay, though improving, is likely to remain inadequate to detect small sporadic thefts. Moreover, since safeguards techniques concentrate on where strategic material is supposed to be, rather than on where it is not supposed to be, it is hard to tell where stolen material has gone. Recovery presents exptreme difficulties, especially after a theft from the vulnerable transport network[11] (tens or hundreds of shipments daily in a mature ^{239}Pu economy). Theft anywhere in the world endangers all countries, however efficient their domestic safeguards. The strategic sanctions which inhibit

[9] Holdren, J P, "Radioactive Pollution of the Environment by the Nuclear Fuel Cycle", 23d Pugwash Conference, Aulanko, 30 August—4 September 1973; *Bull Atom Scient 30,* 8, 14 (1974).

[10] US General Accounting Office B-164105 (7 November 1973); Rosenbaum, D M *et al,* "A Special Safeguards Study", USAEC, 1974, reprinted at 120 *Congr Rec* S6623-6 (30 April 1974).

[11] The recent USAEC decision that substantial shipments will be accompanied by armed guards will presumably make the shipments more conspicuous. It is doubtful that such measures will deter any but the casually curious.

international nuclear aggression today are useless against anonymous and unattributable threats.

Safeguards more costly and thorough than those now applied to fissile isotopes have failed to halt aircraft hijackings, bank robberies, and the black market in heroin (whose black-market price is comparable to the open-market price of ^{239}Pu; nobody knows, one hopes, what ^{239}Pu might cost in a demand-stimulated black market). Such analogies suggest, and the technical literature tends to confirm, that it is impossible to prevent the theft of strategic material by sufficiently determined groups whose motives "are subversive or economic"[12].

b) The marine carriage of liquefied natural gas (LNG) is now in its infancy but has a current doubling time under two years[13]. Receiving terminals for large and highly sophisticated LNG tankers are planned or being built in urban and semi-urban harbours[14]. LNG is essentially liquid methane (CH_4), carried at about −165°C. If spilled on water, it boils extremely quickly, forming a cloud of methane gas which, by reason of its extreme cold, is denser than air and forms a ground-level cloud or plume. Such a plume—which might, in the case of a large tanker spill, extend at least

[12] Geesaman, D P, "Plutonium Diversion", presented at Energy Panel on Radiological Issues Related to Nuclear Power Plants, Science & Technology Council, California State Assembly, 15 June 1972. The between-the-lines logic of the masterly Willrich-Taylor survey[6] leads one to surmise that the steps for transition to a fission-free economy should include consumption of all transuranic stocks in low-power-density, remotely sited, and diligently guarded reactors. Efficient and perpetual control of such stocks would otherwise require "very strict police control of the entire world. . . .This will be difficult to achieve and does not lead us to a very attractive future society." (Alfvén, H, *Bull Atom Scient 30,* 1, 4 (1974).

[13] Faridany, E, "LNG: Marine Operations and Market Prospects for Liquefied Natural Gas 1972-1990", QER Special no. 12, Economist Intelligence Unit, London, 1972.

[14] Fay, J A and MacKenzie, J J, *Environment 14,* 9, 22 (1972).

5000 m downwind within 10-20 minutes[15]—is asphyxiating. It is readily ignitable, and will burn to completion in a turbulent diffusion flame reminiscent of the 1937 *Hindenberg* disaster (but 100× as big). Spillage of LNG after a collision or hard grounding (or, conceivably, after sabotage of a tanker or terminal) would be promoted by brittle fracture[16, 17] of metal structures, such as hulls, on impact of the cryogenic cargo; experience of such incidents suggests that plate failure could propagate and cause an LNG carrier to unzip like a banana. The energy content of a single LNG carrier of 125,000 m^3 (the standard size now being built) is equivalent to approximately 55 Hiroshima bombs.

c) A recent paper[18] plausibly suggests that a large spill of oil in the Beaufort Sea is physically capable of reducing the albedo (reflectivity) of large areas of the Arctic pack ice. The oil could emulsify into small, highly persistent droplets which, circulated in the currents, would collect on the underside of the pack ice. Seasonal melting on top and freezing on the bottom would then yield a darkened ice surface in a few years. It is known[19] that the Arctic pack ice is in an extremely delicate state and could be quickly melted by certain types of small perturbations, with profound and irreversible effects on world climate. A small

[15] Williams, Lt Cmdr H D, *Proc Marine Safety Council* (US Coast Guard) *28*, 9, 162 (1971) and *29*, 10, 203 (1972). The similar estimates given by Fay and MacKenzie[14] have a somewhat firmer theoretical base and are probably underestimates rather than exaggerations. Naturally, nobody has yet tried a full-scale experiment.

[16] Thomas, W duB and Schwendtner, A H, "LNG Carriers: The Current State of the Art", Paper 13, LNG/LPG Conference, London, 21-22 March 1972, *Shipbuilding and Shipping Record:* also in *Trans Soc Naval Archit Marine Eng* 79:440 (1971).

[17] Dobson, R J C, "Problems in the design and construction of liquefied gas carriers", Paper 7, LNG/LPG Conference, *op. cit.*

[18] Campbell, W J and Martin, S, *Science 181:*56 (1973).

[19] Study of Man's Impact on Climate (SMIC), *Inadvertent Climate Modification,* MIT Press (Cambridge, Massachusetts), 1971.

reduction in ice albedo, especially during April and May, could promote melting very effectively[20].

Risks of this magnitude—all uninsured and uninsurable—clearly deserve a more earnest public debate than they have received. Risk (a) is the subject of intense professional worry and vague public unease, though it has become more visible during 1974; (b) has stimulated a small, comparatively obscure literature and a tendency for the responsible agencies to look the other way[21]; and (c) is virtually unknown except to a handful of climatologists, who are taking it very seriously. It is worth noting further that all three of these risks are acute, not chronic: nobody knows the cumulative long-term effects of modern energy technologies on *e.g.* human health, genetic resources, the stability of ecosystems, or many other forms of biological capital. Interference with natural energy flows by means of intensive agriculture offers a particularly worrying example. Indeed, it is hard to think of any current energy technology in extensive use that does not hold the potential for serious long-term environmental risks—risks which may today be wholly unsuspected.

Ethics

The strong ethical ingredient of future-risk problems is perhaps most clearly seen in the perennial controversy, apparently as far from resolution as ever, over the rather arbitrary standards for human exposure to ionizing radiation. Critics of these standards argue, rightly, that safety has not been proved. Advocates reply, correctly but perhaps immorally, that harm has not been proved. Critics point to disquieting statistical trends—inconclusive but suggestive—and say that the burden of proof should rest on the innovator, as is theoretically the case with even the most

[20] *Proc 24th Alaskan Sci Conf* (Fairbanks), 15-17 August 1973, *passim.*

[21] Some agencies, however, are worried. US East Coast landings of LNG may prove hard to license, and some major insurance underwriters are also becoming wary of large LNG tankers and terminals.

beneficial drugs. Advocates can take comfort in the difficulty of demonstrating any real low-level effects, owing to the lack of controls in large populations and the lack of statistical significance of small long-term effects in small populations. And while several billion people are being asked to sign by proxy an irrevocable Faustian bargain[22, 23] whose small print they have not read and are told they cannot understand, those perhaps most affected—thousands of generations yet unborn—are not consulted at all. As Kneese stresses[23], this sort of ethical issue is beyond technical scope or competence, is too important to be left to experts, and requires instead to be resolved *explicitly* in the most open fora of society. This will be slow and hard, but is hardly a dispensable luxury. Low-probability but high-consequence risks, risks of unknowable size, and risks whose avoidance demands social stability into the indefinite future may be impossible to assess without social innovations and ethical conventions.

Conflict

Issues as complex, but of a mixed ethical and pragmatic character, are raised by the enormous conflict potential of inequities in the distribution of world fuel resources: risks of local disaster through nuclear accident are irrelevant if we create global disaster through nuclear war. The inherent delicacy of Middle Eastern politics, the likelihood of massive purchases of armaments and of further nuclear proliferation there, the prospect of increased energy competition between the USA and other OECD countries, and the growing importance (strategically and for the strength of domestic currencies) of such

[22] Weinberg, A M, *Science 177:*27 (1972) and *Nuclear News,* Dec 1971, p 33.

[23] Kneese, A V, "The Faustian Bargain", *Resources* (Resources for the Future, Inc, Washington, DC, September 1973); Edsall, J T, *Environmental Conservation I,* 1, 32 (1974).

commodities as Siberian gas, Canadian tar sands, and American coal and enriched uranium need no emphasis. The divergent interests now tugging at physical resources, too, will become more and more concerned with intangible resources, with know-how, as the nations best able to afford difficult research monopolize the results to their own advantage and ignore or exploit the technological dependence of others. Such advances as would most benefit the poorer countries, however, they will probably have to make for themselves as best they can, since the technically advanced countries are for the most part too busy with their own problems to consider the workings of an energy-conserving, labour-intensive, decentralized society.

11

Outer Limits

OUTER LIMITS

— 11 —

The ultimate constraints—local, regional, and global—on human energy conversion raise important medium- and long-term issues that would defy resolution by existing institutions if, through fortuitous evasion of the many types of inner limits described earlier, the outer limits should become of more than academic interest. We shall not consider here possible outer limits that have yet to be even roughly quantified, *e.g.* those imposed by the availability of land or of skilled labour or of total capital (over $\$10^{12}$ needed for the world oil industry alone by the mid-1980s) or by the biological implications of byproducts such as transuranic isotopes or sulphur compounds. We shall concentrate on climatic limits, where the current state of knowledge is as follows:

Manmade particulates, some from combustion, already form a substantial fraction of the atmospheric load. Combustion (except of hydrogen) also causes an upward trend, averaging 0.2%/yr, in the atmospheric concentration of carbon dioxide. These two effects are in some respects competitive in their effects on global climate, but may in other and perhaps more important respects be coöperative, tending to reinforce the effects of manmade heat. They may become important in the first half of the next century. Some scientists[1] suspect they may already be significant on a regional scale: this case is neither proved nor disproved.

The influence of manmade heat—the end product[2] of any

[1] Bryson, R A, *Ecologist 3,* 10, 366 (October 1973) and *Science 184:*753 (1974).

[2] We refer here not to local "thermal pollution"—the discharge, often ecologically important, of "waste" heat from a power cycle—but to the requirement of the Second Law of Thermodynamics that *all* energy end as low-temperature heat, no matter for what purpose or by what means it is converted.

energy conversion, regardless of the technology used—is already great locally and will soon become significant regionally. Most large industrial areas, including several with areas of order 10^{10} m^2, already add of the order of 10% of net solar heat input to the surface; such areas are projected to intensify, multiply, spread, and merge. The distribution of such regional "heat islands" will be very uneven and mainly at high northern latitudes. The manmade power density of Manhattan Island in New York City (over 700 W/m^2 from man, compared with 93 net from the sun) is very exceptional[3], but perhaps illustrates what can be done.

Global climate is capable of rapid transitions between very different states. These transitions can be triggered by relatively small perturbations applied at particularly sensitive leverage points such as the floating Arctic pack ice[3]. Manmade heat might well be able to provide such a perturbation, at present growth rates, in the first half of the next century, and could certainly do so shortly thereafter[4]. An immense amount of intricate research must be done to determine when such a threat might arise, since it must not be studied empirically. Such research might not yield a definite answer. Though from present, very limited, climatological knowledge it appears that the present pattern of energy growth probably does not present gross climatic (as opposed to biological) risks over the next two or so decades, alternative long-term strategies not involving energy-intensive technologies must be developed promptly in case, as is plausible, global thermal limits make it inadvisable to seek an energy increase of factors of ten, or in case, as is virtually certain, other emission limits impose an even more stringent ceiling on combustion of fossil fuels. (For example, even on the most sanguine view of CO_2 constraints, the use of coal will not be resource-limited.) Any long-range commitment now to an order-of-magnitude energy increase is definitely premature[4];

[3] Study of Man's Impact on Climate (SMIC), *Inadvertent Climate Modification,* MIT Press (Cambridge, Massachusetts), 1971.

[4] Lovins, A B, "Thermal Limits to World Energy-Use", *Bull Atom Scient* (to be published).

within the next decade or so, improved knowledge of the effects of combustion products (*supra*) could be found to impose limits considerably earlier.

Regional climate is capable of periods of prolonged aberration from the mean, such as repeated seasons of drought or storm. Such fluctuations have profound effects on social and economic systems, especially in the present precarious state of world food supplies[5]. There is at present no firm evidence to implicate in or to exonerate from regional climatic disturbances possible manmade perturbations[1]. Urgent research into this question, and into the influence of climatic change on man's activities, is now beginning.

The many unresolved questions about basic climatic mechanisms make present rapid energy growth rates disquieting, and suggest to many climatologists that a policy of caution should be adopted without delay. These doubts imply also that world energy problems cannot be realistically treated as a sum of national problems in a strongly interactive and nonlinear world. International institutions able to regulate man's influence on climate may become essential soon, especially in view of major water and deforestation proposals now under study.

[5] Lovins, A B, "Long-Term Constraints on Human Activity", *Growth and its Implications for the Future,* part 2, pp 1251-67, Serial 93-28, Subcommittee on Fisheries and Wildlife Conservation and the Environment, Committee on Merchant Marine and Fisheries, US House of Representatives, May 1974 (USGPO, 1974).

12

Conclusions

CONCLUSIONS

— 12 —

This paper has only scratched the surface of a body of problems whose detailed exposition would be the work of many lifetimes; yet even at this level of detail, the information density has been so high that it would be quite wrong to attempt a summary. It may be useful to present instead certain inferences which the author believes could properly be drawn from the foregoing assessments. To sharpen the focus, the number of such conclusions has been arbitrarily limited to ten.

a) The rapid energy growth rates that most industrial countries have long maintained cannot continue for much longer. Governments should adjust to this reality, and should devise long-range strategies consistent with it and with the other resource constraints that it entails. Even our ability to maintain current *levels* of per-capita energy conversion in many rich countries over the next few decades is in doubt.

b) Most technical fixes that increase energy supply are slow, costly, risky, and of temporary benefit; most social or technical fixes that reduce energy demand are fairly quick, free or cheap, safe, and of permanent benefit. Industrial countries should immediately undertake lasting and fundamental (not merely temporary and cosmetic) measures to conserve energy in all sectors and forms, and particularly to minimize the consumption of oil and natural gas. The methods that have so effectively promoted these forms of consumption must now be put into reverse. Political imagination and strong moral leadership will be required[1].

[1] A special burden in the many countries which have automobile industries with immense economic and political power will be devising ways to divert those industries' human and technical resources to production of mass-transit vehicles, more durable and less energy-intensive cars, and "soft" energy hardware (wind and solar systems, heat pumps, total energy systems, etc). It is salutary to recall that in World War II the auto industry was completely reconverted in 6-9 months.

c) Means of minimizing the social and environmental costs of mining, converting, and burning coal are urgently needed, as coal and its synthetic products offer many countries their only short- and medium-term bridge from a petroleum economy to a sustainable energy economy. The social and technical problems of coal are substantial but can be adequately solved[2].

d) A diverse range of "unconventional" energy technologies, especially those based on energy income rather than on energy capital, should be developed and deployed as quickly as possible with the help of suitable incentives. Both new and present energy technologies should be integrated into systems that use each form of energy as appropriately and efficiently as possible.

e) The fiscal and human resources now devoted to nuclear fission programmes, particularly to fast breeder reactor development, should be directed forthwith to aims (b)-(d). Governments should suspend their nuclear programmes until enough infallible people can be found to operate them for the next few hundred thousand years[3] and until all those affected by such programmes have been consulted (which may present technical difficulties). Failing all this, contingency planning should at least include a fission-free option which should be taken seriously at policy level: if, as many

[2] One might make a reservation about the sub-micron particle problem (*supra,* p 41); but presumably that can be evaded through coal conversion to synthetics. This problem is largely inseparable from that of heavy metals in fossil fuels; removal of trace heavy metals might also require coal conversion, but does not appear in principle to require exotic technologies.

[3] It may be argued that we cannot do without fission; but (a) we are in effect doing without it now, (b) we have at least two long-lived major technologies—coal and solar—that we *know* will work and that have no problems remotely comparable to the safety, waste, and safeguards problems of fission, and (c) if we really cannot do without fission, then—assuming our present engineering problems with it to be overcome—all long-term planning becomes academic.

experts expect, there is a major reactor accident or safeguards failure in the next 10-15 years, fission may be politically impracticable.

f) All oil and gas resources should be carefully husbanded—*i.e.* extracted as late and as slowly as possible. Our descendents will be grateful. We, too, shall need a long bridge to the future.

g) Governments and their constituencies in rich countries should begin to contemplate seriously and to decide upon the changes in lifestyles that energetic and other constraints will soon impose—changes that may well be desirable on other grounds[4].

h) The energy technologies most appropriate to the poorer countries are also likely to be those on which the richer countries will mainly rely in the long run. These technologies (requiring relatively little capital or infrastructure) should therefore be developed and transferred—with careful attention to the corresponding need for social and institutional transfer (which may have undesirable side-effects)—even though that may mean advancing the development schedule that the rich countries' own short-term needs would dictate. World trade patterns should then be realigned to take account of the redistribution of energy. Interim means should be urgently sought to assist those poor countries whose importation of energy-intensive technologies (especially in agriculture) has made them dependent on energy which, due to others' extravagance, they can no longer afford.

[4] Many, perhaps most, of these changes could be for the better. Nonetheless, very careful planning will be needed to minimize discontinuities. The transition to a sustainable energy economy has much in common with a transition to a macro-stable, micro-variable economy of stock[5] that prefers services to material consumption. In both cases, stabilization or reduction of population size and local self-sufficiency in food production must be top priorities.

[5] Daly, H E, *Toward a Steady-State Economy,* W H Freeman (San Francisco), 1973.

i) Countries not yet industrialized should make every effort to avoid others' mistakes by (in general) minimizing their populations, maximizing their agricultural self-sufficiency (particularly in the use of sustainable, labour-intensive methods), and minimizing their need for large amounts of industrial energy. Industrial countries should do the same. The rapid introduction of diffuse and medium-scale solar and wind power, and of technologies for converting organic materials into fuels, should help, especially in tropical countries, to reduce urbanization by improving rural economies and living conditions. It is impossible to bring world per-capita energy conversion up to rich-country levels; on the contrary, the latter levels should be considerably reduced as an essential part of redistribution and as a step towards sustainability. Rich countries may find it helpful in the medium- and long-run to export some energy-intensive industries—petrochemicals and nitrogen-fertilizer manufacture to the Persian Gulf, aluminium-smelting to centralized solar installations in the tropics, etc. This could benefit all concerned and could open up important regional options (*e.g.* for Norway to export surplus hydroelectricity in return for food etc).

j) All countries should note the following points of policy:

   1) Economic processes are often assumed to have limited materials inputs, unlimited energy inputs, and market-limited outputs. Boundary conditions on the energy inputs are now needed[6]. In short, physical law imposes limits on technology, and biophysical as well as economic factors must govern energy decisions.

   2) We must devise a science and a technology of energy impact analysis so that we can make energy *a critical variable in all policy decisions,* rather than leaving it to

[6] This is so if only because the capacity of a given volume of space-time safely to accept heat is a *strictly* nonrenewable and nonsubstitutable resource (a concept foreign to most economists).

emerge *de facto* from decisions taken on other grounds. Otherwise we may find ourselves becoming net exporters of energy to the Middle East by bartering energy-intensive goods for oil, or committing other such absurdities. Moreover, we must all learn to *see* energy—not just fuels and electricity, but steaks, paper, bottles, everything. We cannot value a commodity whose presence we do not perceive.

3) We must distinguish need from demand and curb needless demand: concepts of energy as a medium of exchange, and of societal energy allocation by various means, deserve refinement and study. We must reduce or reverse rates of energy growth in industrial societies before it is done for us by the consequences of present rapid growth[7].

4) Economists' perception of energy is likely to improve if they relearn that money is a medium of barter for real physical resources (and labour, water, land, etc) *but it is not itself such a resource:* hence that capital investment in an energy technology is not a substitute for, or a way of evading, expenditures of real physical resources (such as energy), but is rather *a way of shuffling those expenditures to other times and places*—perhaps so remote in the economic system that the expenditures are not perceived as inputs to the technology being capitalized and hence are not deducted from its gross output in computing its net output[8].

[7] A system with rates of change rapid compared with built-in delays is inherently unstable. The instability can be reduced by reducing the rates of change; or by reducing those delays that are due to social rather than to natural causes (*sc* by autocracy); or by accurate long-range predictive planning methods that do not now exist. The world appears to be a system in this state.

[8] Lovins, A B, "Net Energy", *Stockholm Conference Eco IV:* 1, Friends of the Earth (529 Commercial St, San Francisco, California 94111), 1974.

5) In order to make rational long-term decisions—longer than about 20 years[9]—we must learn about the nonlinear behaviour of biological systems, rely less on discounted-cash-flow planning, stop discounting future risks, greatly lengthen our time horizons, rely on economic sensitivity analysis rather than on marginal-cost analysis, and anticipate sociotechnical lags by sophisticated and diverse strategies of research and development. Slower or negative energy growth, properly used, can buy the time we need to do these things.

6) Policy advice should start to come now from those who have tried to prevent our present difficulties, not from those who have caused them but who still dominate energy planning. People who think mainly in terms of econometric extrapolation, marginal mills per kilowatt-hour, economies of scale, and centralized electrification are not the planners who can most constructively address our energy problems. Moreover, *the important issues of energy strategy are not technical and economic but rather social and ethical,* and cannot be properly framed by those whose vision is purely technical: "their belief in the effectiveness of power blinds them to the greater effectiveness of abstaining from its use."[10]

7) We must adopt explicitly the methodology of *decision-making under uncertainty,* recognizing that even events of extremely low probability may occur and must be guarded against. High-technology, highly bureaucratized, highly centralized systems are the most vulnerable to the unexpected. Governments should try to diversify their risk portfolios.

8) Finally, we must keep more clearly in mind the ways in which the time-scale we select for planning can affect our choice of strategies. Pursuing a short-term goal

[9] As Maurice Strong points out, presently used discount rates imply that the present value of a child is essentially zero.

[10] Illich, I, *Energy and Equity* (Calder & Boyars Ltd, 18 Brewer St, London W1R 4AS; 1974).

(such as independence from imported oil) can foreclose long-term options and lead to similar but less tractable problems several decades hence, just as we are severely restricted now by short-term planning in the past. Planning is a process of formal procrastination, and its aim in an uncertain world must be to keep as many options *open* as possible. Present energy policies, however tacit and ill-constructed they may be, are quickly destroying the options that mankind, living and unborn, will need for millenia.

Author

Amory Bloch Lovins resigned a Junior Research Fellowship of Merton College, Oxford in 1971 to become full-time British Representative of Friends of the Earth Inc, working mainly in energy and resource policy. A consultant physicist in the USA and UK since 1965, he is an energy consultant to the Organization for Economic Coöperation and Development (Paris), the United Nations Environment Programme (Nairobi), the International Federation of Institutes for Advanced Study (Stockholm), the Massachusetts Institute of Technology (Cambridge), and other agencies, groups, and projects in several countries. He has testified before Parliamentary and Congressional Committees, has broadcast extensively, and is the author of three earlier books, several monographs, and numerous articles, technical papers, and reviews.